EYEWITNESS
ENERGY

Written by
Dan Green

Cross-section
of the Sun

Paint mixing
in water

Athletes at the
start of a race

Wind turbine

DK | Penguin
Random
House

Consultant Jack Challoner

DK DELHI
Senior editor Bharti Bedi
Senior art editor Nishesh Batnagar
Project editor Priyanka Kharbanda
Project art editor Tanvi Sahu
Assistant editor Sonam Mathur
Assistant art editor Ankit Singh
Senior DTP designers Harish Aggarwal, Neeraj Bhatia
DTP designer Pawan Kumar
Senior picture researcher Sumedha Chopra
Jacket designer Dhirendra Singh
Managing jackets editor Saloni Singh
Managing editor Kingshuk Ghoshal
Managing art editor Govind Mittal
Pre-production manager Balwant Singh
Production manager Pankaj Sharma

DK LONDON
Senior editor Chris Hawkes
Senior art editor Spencer Holbrook
Jacket editor Claire Gell
Senior jacket designer Natalie Godwin
Jacket design development manager Sophia MTT
Producer, pre-production Gillian Reid
Producer Vivienne Yong
Managing editor Linda Esposito
Managing art editor Philip Letsu
Publisher Andrew Macintyre
Associate publishing director Liz Wheeler
Design director Stuart Jackman
Publishing director Jonathan Metcalf

First published in Great Britain in 2016
by Dorling Kindersley Limited
80 Strand, London WC2R 0RL

Copyright © 2016 Dorling Kindersley Limited

A Penguin Random House Company
10 9 8 7 6 5 4 3 2 1
001 – 288034 – June/16

A CIP catalogue record for this book is available
from the British Library.

ISBN 978-0-2412-3560-7

Printed and bound in China

A WORLD OF IDEAS:
SEE ALL THERE IS TO KNOW

www.dk.com

Dominos
falling in
sequence

Kinetic energy
is transferred from
the falling ball to the
stationary balls

Nuclear bomb
explosion

Contents

Prism refracts light

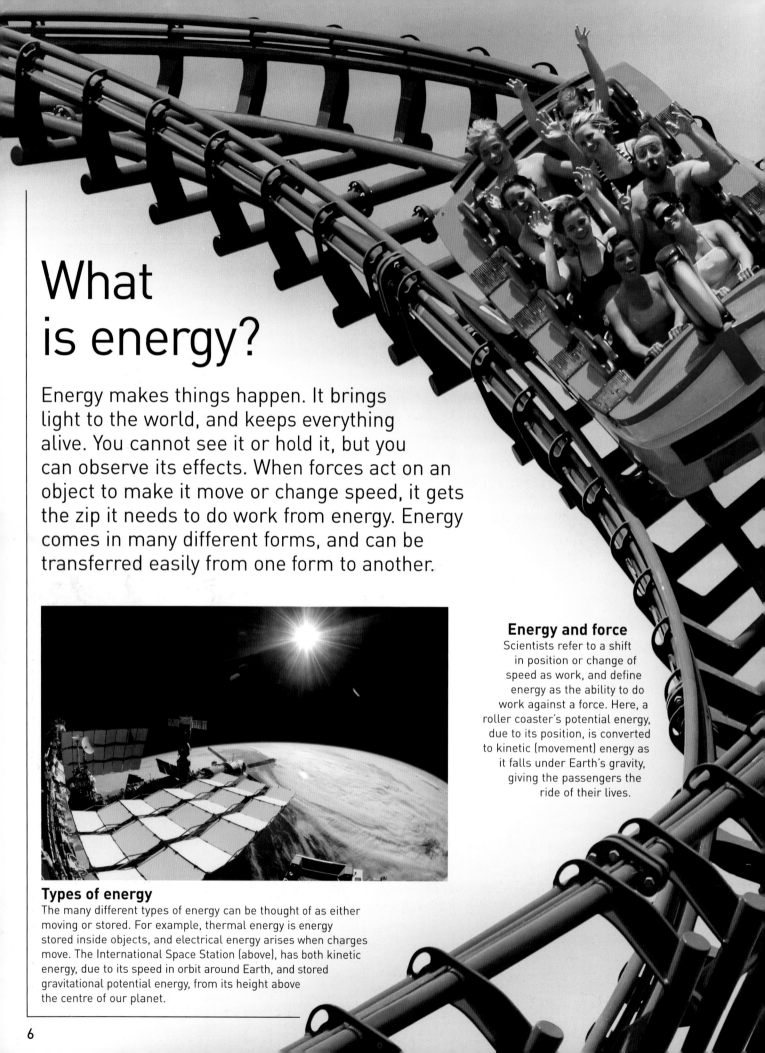

What is energy?

Energy makes things happen. It brings light to the world, and keeps everything alive. You cannot see it or hold it, but you can observe its effects. When forces act on an object to make it move or change speed, it gets the zip it needs to do work from energy. Energy comes in many different forms, and can be transferred easily from one form to another.

Energy and force
Scientists refer to a shift in position or change of speed as work, and define energy as the ability to do work against a force. Here, a roller coaster's potential energy, due to its position, is converted to kinetic (movement) energy as it falls under Earth's gravity, giving the passengers the ride of their lives.

Types of energy
The many different types of energy can be thought of as either moving or stored. For example, thermal energy is energy stored inside objects, and electrical energy arises when charges move. The International Space Station (above), has both kinetic energy, due to its speed in orbit around Earth, and stored gravitational potential energy, from its height above the centre of our planet.

Energy from the Sun

A lot of energy can be harnessed from reactions involving the nucleus (core) of an atom – the basic unit of matter. These are called nuclear reactions. Nuclear reactions inside the Sun flood the Solar System with electromagnetic radiation – heat, visible light, and high-energy radiation. Plants, algae, and some bacteria convert the energy of this electromagnetic radiation – radiant energy – into food, which sustains life on Earth. Uneven heating by the Sun also drives our planet's weather, and the ocean currents that regulate Earth's climate.

Geysers (springs that spurt hot water) in Iceland heat up and erupt frequently

Energy on the move

Although energy moves, it is not a substance. Instead, as the temperature of a material increases, its atoms and molecules move faster, increasing their kinetic energy. It is the internal energy of these tiny vibrating particles that causes substances to change state, which is what happens when water turns to steam on heating.

Wasted energy

Transferring energy allows us to make use of it. Light bulbs illuminating these skyscrapers (right) take electrical energy and convert it into light energy. However, energy is always lost in this process of conversion. In a car, nearly all of the energy fuel produces is lost – as heat in the engine, or as in friction between moving parts.

In a smash, the kinetic energy of motion has crumpled the car, producing sound and heat

Doubling the speed of a moving object increases its kinetic energy by four times

Conservation of energy

Energy is wasted when it is transferred, but not destroyed. When two objects collide, such as a car and a barrier in a crash test, one of them loses energy and the other gains it. However, the total amount of energy remains constant. This is a fact with no exceptions in the Universe.

The Sun

The Sun is the ultimate source of energy for life on Earth. It is a hugely energetic mass of seething gas. The reactions in its core pump out the same amount of energy as that produced by 100 million billion coal-fired power plants. This radiant energy, which includes light and heat, takes just eight minutes to travel to Earth, a distance of 150 million km (93 million miles).

How does the Sun shine?

The Sun shines because it produces huge amounts of light energy in a process called nuclear fusion. In it, nuclei – the central parts of atoms – of hydrogen are joined together to build nuclei of helium, releasing energy in the process. This energy takes thousands of years to travel to the Sun's visible surface.

Huge eruptions of gases, called prominences, loop over the Sun's surface

Cooler sunspots look darker than the surrounding gas

Prolonged exposure to sunlight leads to sunburn, and can cause skin cancer

Nuclear fusion reactions in the core use up 600 million tonnes of hydrogen every second

In the radiative zone, energy is transferred in the form of radiation – energy that travels as electromagnetic waves

The Sun's energy

About half of the energy coming from the Sun is visible light. The other half is mostly infrared radiation, which we feel as the Sun's heat. The Sun also shines with other frequencies of radiant energy, and can be seen in radio waves, ultraviolet, X-rays, and gamma rays. The Earth's atmosphere absorbs most of the harmful ultraviolet rays, but some reach the surface. This is what causes sunburn.

Photosphere is the visible surface of the Sun

Life-giver

The constant supply of energy from the Sun keeps life on Earth going. Every green plant or animal uses a portion of the available energy. Many living things get their energy directly from the Sun, while others take it by consuming other organisms.

Carnivores, such as lions, are secondary consumers, which survive by eating primary consumers

Producers, such as green plants, harness the Sun's energy to make food

Herbivores are primary consumers, and get energy from plants. This group includes large animals, as well as many insects and bugs.

Core's temperature is around 15 million °C (27 million °F)

Energy in the convective zone is transferred via convection to the surface

Light in the sky

High-energy particles streaming off the Sun cause eerie shifting lights in the skies over the northern and southern parts of Earth. Caught by Earth's magnetic field, these particles spiral in towards the poles. When they hit the upper atmosphere, their energy is absorbed by gases, which re-emit it as light energy.

Fade to black

At the end of its life, the Sun will run out of fuel, and life on Earth will grind to a halt. Long before that, the planet will be slowly roasted, as the Sun heats up and swells up to become a red giant star.

An artist's impression of Earth baked by the dying Sun

Sun's outer layer, or corona, streams out into space

Earth's energy

Earth still holds on to the heat energy that was released when it formed around 4.5 billion years ago. The heat from its thick, treacly interior pushes the outer parts of the planet about, throwing up mountain ranges, causing volcanoes and earthquakes, and rearranging the globe's surface. The continuous energy supply from the Sun also stirs up the atmosphere, creating the planet's weather systems.

Drifters

Earth's internal energy causes huge slabs of rigid rock, called tectonic plates, to move around the planet's surface. This process breaks up continents where they separate and forms new ones where they meet. When plates rub against each other, such as at the San Andreas Fault, USA, the energy released causes earthquakes.

Monkey business

Japanese macaques soak in a hot spring in Japan's Koshinetsu region. The water in this spring is heated by the heat from within Earth. Between Earth's core and its crust, temperatures range between 500–900ºC (932–1,650ºF). At places where convection currents bring this energy close enough to the surface, it can be harnessed to heat water and generate electricity.

Electric charge created by collision of rock particles, ash, and ice

Energy allowance

Although clouds reflect some of the Sun's energy back into space, just under three-quarters of it is absorbed by the planet's oceans, land, and atmosphere. Atoms and molecules that have been heated up by this energy then radiate it back into space as infrared radiation.

Magma (liquid rock) erupts onto the surface

Desert areas lose heat quickly at night due to lack of moisture in the air

From outer space

Earth occasionally receives some surprise injections of energy. Chunks of rock from space sometimes collide with it. These impacts can be devastating. About 66 million years ago, a 10-km- (6-mile-) wide asteroid crashed into Earth at Mexico's Yucatán Peninsula, delivering a huge amount of energy – equivalent to 100 trillion tonnes of TNT explosives – and triggering a mass extinction that killed off the dinosaurs.

Cosy blanket

Energy from the Sun heats our planet's surface and its atmosphere. Like anything warm, they emit infrared (heat radiation), so that all the energy they absorb is lost back to space. The atmosphere absorbs some of the infrared emitted by the surface, and the surface absorbs some of the infrared emitted by the atmosphere, helping to keep our planet warm. Burning fossil fuels adds gases that boost this "greenhouse effect", gradually raising the planet's temperature.

Clouds and atmosphere reflect about one-third of heat energy into space

Sunlight

Some heat is reflected by Earth into space

Some heat is reflected by the greenhouse gases to Earth

Greenhouse gases in the atmosphere

Soft on the inside

The deeper you go inside Earth, the hotter it gets. For each kilometre you descend, the temperature increases by 30°C (86°F); at the core, the temperature is around 6,000°C (10,830°F). Some of this heat energy is leftover from the planet's birth. However, much of it comes from the radioactive decay of atoms inside the mantle and core. Heat from deep within Earth's mantle can often be transferred to the surface by volcanoes, such as Anak Krakatau, Indonesia (left).

Nature's fury

Uneven heating of the atmosphere causes pressure differences, which power Earth's weather systems. Energy from the Sun absorbed by the oceans also drives currents that transport heat around the planet. Furious storms brew when tropical seas become unseasonably warm, and these can throw out colossal amounts of energy.

Potential and kinetic energy

Energy gets things done – you must have energy if you want to do work. In scientific terms, work is done on an object when it moves in the direction of a force. An object may possess two key types of energy: potential energy, which is energy due to position; and kinetic energy – energy due to motion.

Gravitational potential energy

Potential energy is related to position. It requires work to move something against a force. Lifting this wrecking ball against the force of gravity gives it gravitational potential energy.

Bullet has passed through apple at a speed of nearly 1 km (0.6 miles) per second

To do work

Chemical energy from food enables this man to lift the axe against Earth's gravitational pull, giving it gravitational potential energy. Potential energy is pent-up energy, primed and ready to make things happen. It can take many forms, from gravitational and elastic potential energy, to electric and magnetic potential energy, and even nuclear energy.

Elastic potential energy is stored in the sprung bow

Elastic potential energy

Pull back the string of a bow, wind up a coiled spring in a clockwork toy, or stretch the elastic band of a catapult, and we all know what happens next. Elastic potential energy results from forces changing an object's shape – normally stretching or squashing. Many simple machines are powered by this type of energy.

Energy on the move

All moving objects have kinetic energy. Forces put things in motion, and this movement often happens when potential energy is released or converted. A bobsleigh falls downhill under its own weight. As it speeds down the run, its gravitational potential energy changes into kinetic energy.

Thin metal runners reduce friction between bobsleigh and ice on track

Aerodynamic shape helps reduce air drag and boost speed

Energy passed to the apple is enough to tear it apart in a few milliseconds

Speed and energy

Fired from a high-velocity rifle, this bullet has enough kinetic energy to destroy the apple. The kinetic energy of a moving object depends on both its mass and speed, but it increases in greater proportion to the speed. When something doubles its pace, its energy increases four times over.

The bullet has a lot of kinetic energy because of its high speed

Kinetic energy is transferred from the falling ball to the stationary balls

Energy transfer

In "elastic" collisions between hard steel balls (right), kinetic energy continues to power the balls after the initial collision, and none of that energy is lost. In an "inelastic" collision, some kinetic energy is lost as sound – making the noise you hear when the balls knock together.

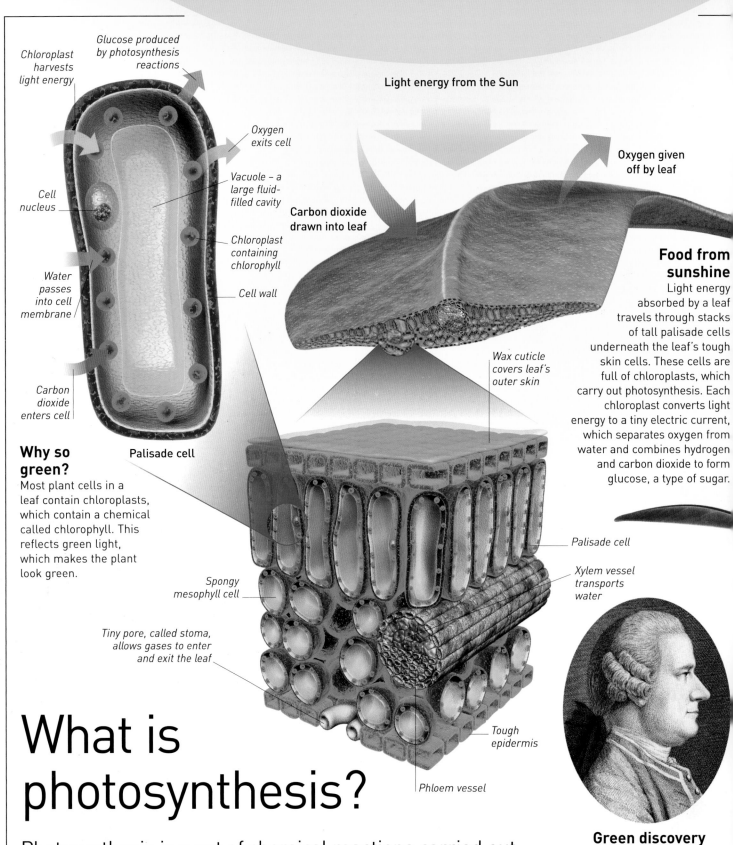

Chloroplast harvests light energy

Glucose produced by photosynthesis reactions

Oxygen exits cell

Vacuole – a large fluid-filled cavity

Cell nucleus

Chloroplast containing chlorophyll

Cell wall

Water passes into cell membrane

Carbon dioxide enters cell

Light energy from the Sun

Oxygen given off by leaf

Carbon dioxide drawn into leaf

Food from sunshine
Light energy absorbed by a leaf travels through stacks of tall palisade cells underneath the leaf's tough skin cells. These cells are full of chloroplasts, which carry out photosynthesis. Each chloroplast converts light energy to a tiny electric current, which separates oxygen from water and combines hydrogen and carbon dioxide to form glucose, a type of sugar.

Wax cuticle covers leaf's outer skin

Why so green?
Most plant cells in a leaf contain chloroplasts, which contain a chemical called chlorophyll. This reflects green light, which makes the plant look green.

Palisade cell

Palisade cell

Xylem vessel transports water

Spongy mesophyll cell

Tiny pore, called stoma, allows gases to enter and exit the leaf

Tough epidermis

What is photosynthesis?

Phloem vessel

Photosynthesis is a set of chemical reactions carried out by green plants, microscopic plankton, and some bacteria and algae, to absorb sunlight and build their bodies. Photosynthesizing organisms provide food for all living things on our planet that do not take energy directly from sunlight, and fill Earth's atmosphere with life-giving oxygen.

Green discovery
In 1779, Dutch scientist Jan Ingenhousz (above) discovered that only the green parts of plants release oxygen, and only when they are in sunlight. He put plants under water and watched how the leaves produced bubbles of gas in sunlight.

Oxygen factories

The chemical reactions in the microscopic cells of photosynthesizing organisms affect the make-up of the planet's air. Billions of years ago, the atmosphere contained a suffocating mixture of toxic gases. However, when the first photosynthesizers evolved, they removed much of the carbon dioxide from the atmosphere and pumped oxygen back into it. Trees and other plants in rainforests, such as this one in the Danum Valley, Malaysia, produce large amounts of oxygen.

Plant uses sugary glucose to build its body

Cyanobacteria

Microscopic bacteria called cyanobacteria produce almost one-third of all the oxygen in our atmosphere. Cyanobacteria are some of the oldest living things on Earth. They are found in every place where there is light, even in extreme environments, as floating plankton in the oceans, and sometimes as colonies, such as these stromatolites in Shark Bay, Australia.

Roots draw up water from the soil

Future fuel

Because plants store the light energy from the Sun as chemical energy, they can be burned to release this energy. This stored energy makes fossil fuels powerful energy sources. However, burning fossil fuels releases the harmful carbon dioxide that the plants had locked up while alive. Instead, imagine that we could harvest the limitless energy streaming in from the sky, just like plants. Scientists are working to build artificial leaves that split water to release hydrogen and oxygen.

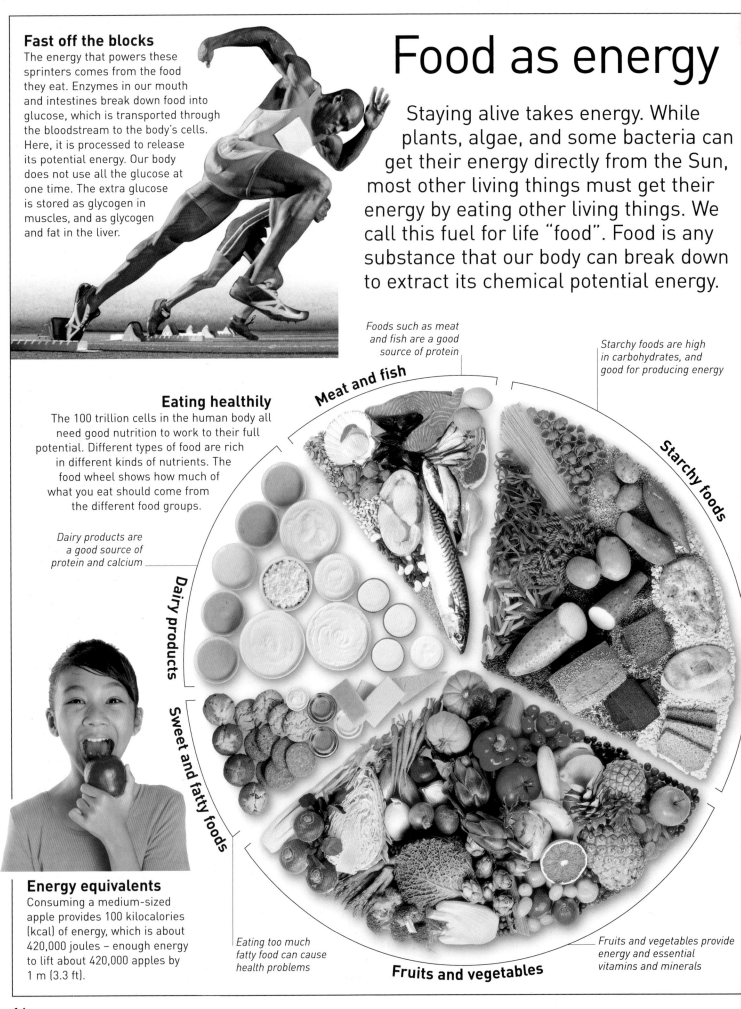

Fast off the blocks

The energy that powers these sprinters comes from the food they eat. Enzymes in our mouth and intestines break down food into glucose, which is transported through the bloodstream to the body's cells. Here, it is processed to release its potential energy. Our body does not use all the glucose at one time. The extra glucose is stored as glycogen in muscles, and as glycogen and fat in the liver.

Food as energy

Staying alive takes energy. While plants, algae, and some bacteria can get their energy directly from the Sun, most other living things must get their energy by eating other living things. We call this fuel for life "food". Food is any substance that our body can break down to extract its chemical potential energy.

Eating healthily

The 100 trillion cells in the human body all need good nutrition to work to their full potential. Different types of food are rich in different kinds of nutrients. The food wheel shows how much of what you eat should come from the different food groups.

Dairy products are a good source of protein and calcium

Foods such as meat and fish are a good source of protein

Starchy foods are high in carbohydrates, and good for producing energy

Meat and fish

Starchy foods

Dairy products

Sweet and fatty foods

Fruits and vegetables

Energy equivalents

Consuming a medium-sized apple provides 100 kilocalories (kcal) of energy, which is about 420,000 joules – enough energy to lift about 420,000 apples by 1 m (3.3 ft).

Eating too much fatty food can cause health problems

Fruits and vegetables provide energy and essential vitamins and minerals

Salmon jumps out of water while swimming upstream

Adult brown bear tries to catch fatty salmon

Fat reserves

If "energy in" is greater than "energy out", the excess energy is stored as fat. Bears must gain weight in the autumn to build up a fat reserve that will last them through the winter hibernation. However, for humans, a layer of fat under the skin and around the organs merely stores up health problems for the future.

High-calorie foods, such as this cake, can provide more energy than low-calorie foods

Measuring energy in food

The energy content of foods is measured in kilocalories (1 kcal = 1,000 calories). Controlling your energy intake is important for maintaining energy levels, as well as a healthy weight. Staying fit helps burn off the calories taken in the diet.

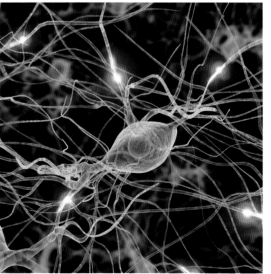

Energy in the body

Your body derives many forms of energy from food. For example, the nervous system uses electrical energy to release tiny jolts of chemical energy, which transport signals to and from the brain. Nerve fibres carrying electric charges (left) inform the brain about what is going on outside the body, and tell the body parts how to move. The brain consumes up to one-fifth of the energy you take in from your food.

Heat

Heat is a process by which energy flows from one thing to another. The particles of a substance are in constant motion and possess kinetic energy. Energy from heat (thermal energy) is a form of kinetic energy possessed by vibrating particles. The more they vibrate, the hotter a material becomes.

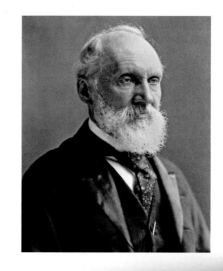

One-way journey
British physicist William Thomson (left), also known as Baron Kelvin of Largs, realized heat only ever travels in one direction, always passing from hot things to cold things. Left to themselves, things lose energy until they are the same temperature as their surroundings. They never get hotter, unless energy is added to them by heating.

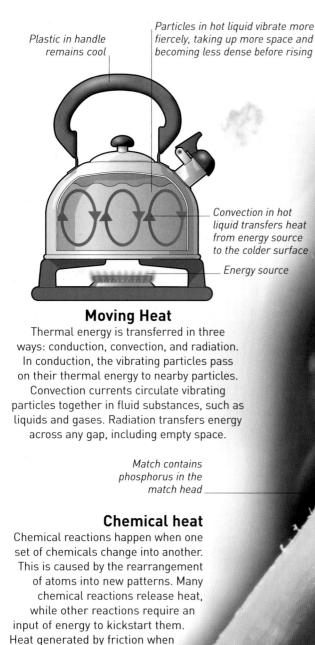

Plastic in handle remains cool

Particles in hot liquid vibrate more fiercely, taking up more space and becoming less dense before rising

Convection in hot liquid transfers heat from energy source to the colder surface

Energy source

Moving Heat
Thermal energy is transferred in three ways: conduction, convection, and radiation. In conduction, the vibrating particles pass on their thermal energy to nearby particles. Convection currents circulate vibrating particles together in fluid substances, such as liquids and gases. Radiation transfers energy across any gap, including empty space.

Match contains phosphorus in the match head

Chemical heat
Chemical reactions happen when one set of chemicals change into another. This is caused by the rearrangement of atoms into new patterns. Many chemical reactions release heat, while other reactions require an input of energy to kickstart them. Heat generated by friction when a match is struck begins a reaction that sets the matchstick alight.

Going through phases

Heat gives the atoms and molecules inside an object energy to vibrate more vigorously. This is why heating substances makes them expand. Adding heat does not always increase the temperature. It sometimes makes the substances change, melting solid ice into liquid water, for instance. The heat energy allows the molecules (in ice) that are packed together tightly to loosen up, changing the state without an increase in temperature.

Icicles melt into water because of heat from the Sun

Hot stuff

When we measure temperature, we are measuring the average kinetic energy of the vibrating atoms and molecules in the object. Thermometers are used to measure temperature. The first accurate thermometer (with a clearly defined scale) was invented by German physicist Daniel Gabriel Fahrenheit in 1714.

Striking a match converts red phosphorus into white phosphorus, which bursts into flame

Friction between tyres and track

There's the rub

When two surfaces rub against each other, the force that acts in an opposite direction to the movement and generates heat is called friction. Friction between the tyres and the road surface provides grip, but also slows the wheels and produces a lot of heat. Friction was key to explaining heat because it generates limitless heat.

Staying warm

The heat you feel on your face from a fire is a type of electromagnetic radiation called infrared radiation. It travels across space and is absorbed by matter, such as your skin, heating it up. In the case of a polar bear, thick fur blocks radiant energy, trapping warmth close to the animal's body.

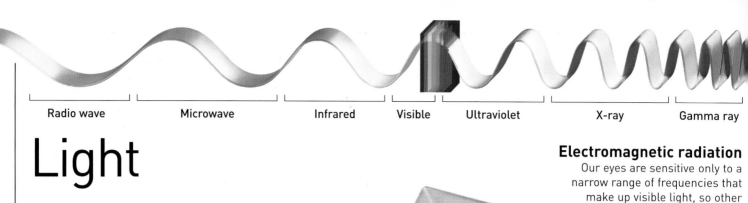

Light

Light is a type of radiant energy. It travels in straight lines through gases, liquids, solid matter, and even the nothingness of space. It behaves as a wave as well as a particle. Frequency describes how fast a wave oscillates. Our eyes are sensitive to the frequencies of visible light, but there are frequencies of light that we cannot see.

White light

Electromagnetic radiation

Our eyes are sensitive only to a narrow range of frequencies that make up visible light, so other forms of electromagnetic radiation are invisible to us. The spectrum of light ranges from low-energy radio waves to high-energy gamma rays, and is known as the electromagnetic spectrum. X-rays and gamma rays have so much energy they can damage cells in our body and cause cancer.

Rainbow colours

Light that is white in colour – such as that from the Sun – is composed of different frequencies of light. We sense each frequency as a different colour. A prism can split white light into a spectrum of colours.

White light, white heat

Hotter objects emit higher frequencies of electromagnetic radiation than colder ones. A cold log emits only invisible infrared radiation, whereas a hot fire glows with visible light. The hotter an object is, the more of the spectrum it emits. If it emits all the colours of the spectrum, it will glow white hot.

Speed of light

At 299,792,458 m (983,571,056 ft) per second, light and other forms of electromagnetic radiation travel faster than anything else. Beams of light rays follow rules that govern how they reflect off surfaces. In 1849, French scientist Hippolyte Fizeau (left) used these properties to get the first accurate determination of the speed of light.

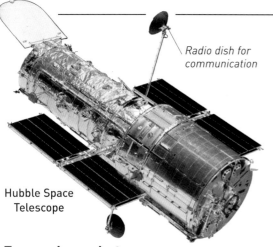

Radio dish for communication

Hubble Space Telescope

Energy in packets

The Hubble Space Telescope (HST) uses an on-board light detector to take pictures of the dimmest objects in the Universe. Light energy can be imagined as travelling as a "blip", or packet, of electromagnetic energy called a photon, which is the basic particle of light. Photons of different frequencies contain different amounts of energy. For example, photons in an X-ray have more energy than photons in infrared radiation. The HST builds images by collecting light, photon by photon, from the most distant galaxies.

Cold light

Many marine animals, bugs, and fungi can create their own light. Jellyfish, molluscs, and some fish living in the dark depths of the deep ocean often house special bacteria in their light-producing organs. These microorganisms can generate light energy via chemical reactions. Barely one-fifth of the energy is converted to thermal energy, so there is no heat, and the light is cold.

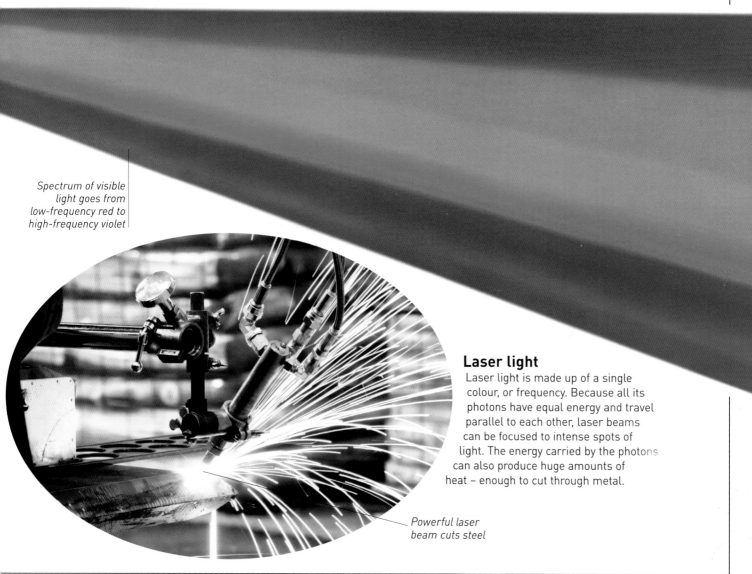

Spectrum of visible light goes from low-frequency red to high-frequency violet

Laser light

Laser light is made up of a single colour, or frequency. Because all its photons have equal energy and travel parallel to each other, laser beams can be focused to intense spots of light. The energy carried by the photons can also produce huge amounts of heat – enough to cut through metal.

Powerful laser beam cuts steel

Sound

Sounds surround us. When a material vibrates, it produces a disturbance in the surrounding atoms and molecules. This kinetic energy moves outwards in ripples and waves, passing on vibrational energy, which we detect as sound. Because noises tell us what is happening around us, most human beings are very sensitive to them.

Wave of compaction

Wave of rarefaction

Good vibrations
Sound travels as vibrations within a solid, liquid, or gas. It moves in waves of compaction (in which particles are closer together) and rarefaction (in which they are further apart). Sound cannot travel through a vacuum, as a vacuum has no particles through which the energy can pass.

Coloured paint visually demonstrates the energy of sound waves

Movement of the cone sends paint leaping into the air

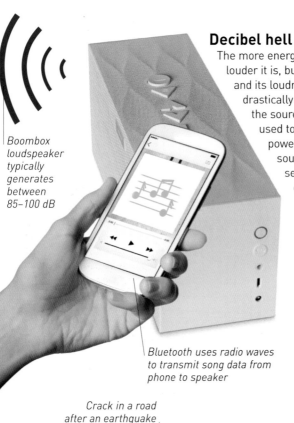

Boombox loudspeaker typically generates between 85–100 dB

Decibel hell
The more energy a sound has, the louder it is, but the sound's energy, and its loudness, decreases drastically with distance from the source. Decibels (dB) are used to measure "received power" – the energy a sound delivers per second at a distance of 1 m (3 ft) from a source. Sound becomes painful at 115 dB.

Bluetooth uses radio waves to transmit song data from phone to speaker

Crack in a road after an earthquake

Telltale tremors
When the rocks of Earth's crust move, the release of stored elastic energy sends huge shock waves juddering through the planet. The first of these are giant sound waves, which rock the surface as earthquakes. Using seismographs, which can detect the energy of these sound waves from thousands of miles away as tiny vibrations of the ground, scientists can map the interior structure of Earth.

Sonic shrimp

The tiger pistol shrimp is the loudest living animal. The shrimp's claw shoots a jet of water that creates an air bubble. This collapses with a bang that tops 200 dB – louder than a passing jet.

Claw cocked like a pistol

Detecting sounds

Ears work like loudspeakers in reverse. Sound waves hit the eardrum, which transmits the energy as vibrations into the inner ear. The cochlea converts the sound waves into electrical nerve impulses, which travel to the brain. The brain interprets the signals as sound.

Ossicles (tiny bones) transfer sound from eardrum to inner ear

Cochlea

Outer parts of the ear direct sounds into the ear canal

Making music

Loudspeakers convert electrical energy into kinetic energy. When music is playing, the loudspeaker's cone bounces in and out. The cone's movement sets up pressure waves in the air, sending sound energy blasting out of the speaker.

Outer ear canal

Eardrum

Cross-section of a human ear

Loudspeaker with paint on cone

Echo back from surface

Seeing with sound

Sonar uses sound energy to map things that are difficult to see. By sending out a signal and listening for the "ping" of the returning echo, the distance to objects can be gauged. Echo sounding maps the ocean floor in detail, and fishing trawlers use it to locate shoals of fish. Ultrasound is used to make images of babies inside the bodies of pregnant women.

Sonar waves from submarine

Paper or plastic cone

Metal cage protects delicate cone

Soundless

The anechoic chamber at the Orfield Labs in Minnesota, USA, is the quietest place on Earth. This specially built room absorbs 99.9 per cent of any sound energy created in it. It is used to design and test products, such as quieter motorbikes.

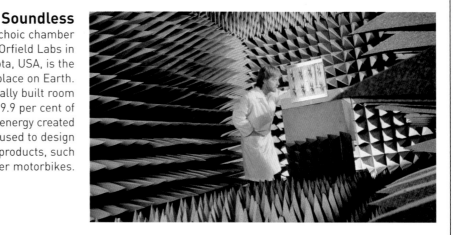

Strong permanent magnet circles coil of wire

Gravitational potential energy

Gravity is a force of attraction between objects with mass. It stops the gases of Earth's atmosphere from drifting off into space, and keeps our feet on the ground. When we raise an object, it gains gravitational potential energy. The higher it is raised, the more gravitational potential energy it gains.

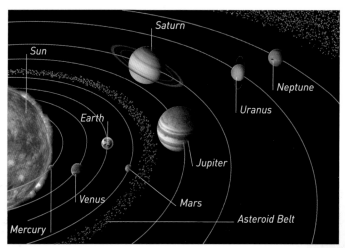

An artist's impression of the Solar System (not to scale)

Forever falling

The Sun's gravity keeps all the planets of the Solar System in motion. However, rather than falling into the Sun, the kinetic energy of the planets keeps them in their orbits. Each planet has gravitational potential energy because of the gravitational pull of the Sun. A planet's gravitational potential energy increases as it moves away from the Sun in its orbit, while its kinetic energy decreases. The opposite is true as the planet nears the Sun.

Defeating gravity

It takes an enormous amount of energy to lift a spacecraft up above the atmosphere. In order to go into orbit around Earth, a spacecraft must travel at more than 8 km (5 miles) per second, or else it will fall back down. This is achieved by huge rocket engines that burn high-energy fuel rapidly. To escape the gravity of a planet or moon altogether, a rocket must achieve escape velocity – for Earth, this is 11 km (7 miles) per second.

Mass matters

At the top of a half-pipe skiing run, a skier trades in his kinetic energy for gravitational potential energy, to get "big air"– leave the ground – and pull off a stunt. However, his kinetic energy will only take him so far into the air before the force of gravity begins to pull him back to Earth. At this point, he cashes his gravitational potential energy back into kinetic energy.

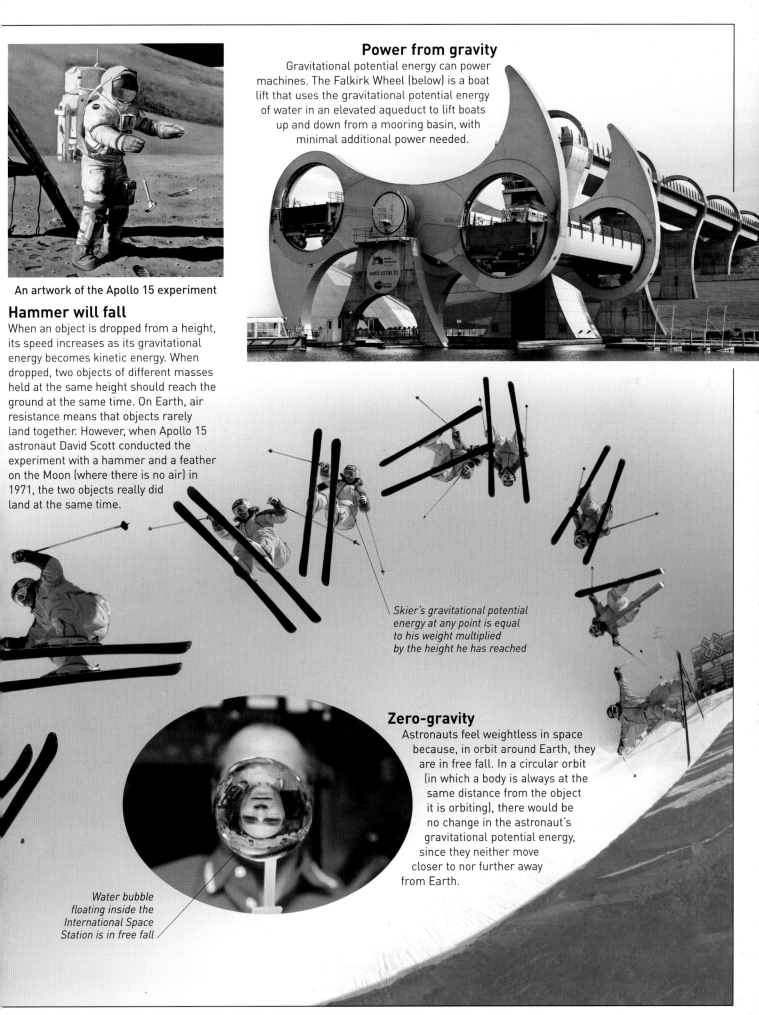

Power from gravity

Gravitational potential energy can power machines. The Falkirk Wheel (below) is a boat lift that uses the gravitational potential energy of water in an elevated aqueduct to lift boats up and down from a mooring basin, with minimal additional power needed.

An artwork of the Apollo 15 experiment

Hammer will fall

When an object is dropped from a height, its speed increases as its gravitational energy becomes kinetic energy. When dropped, two objects of different masses held at the same height should reach the ground at the same time. On Earth, air resistance means that objects rarely land together. However, when Apollo 15 astronaut David Scott conducted the experiment with a hammer and a feather on the Moon (where there is no air) in 1971, the two objects really did land at the same time.

Skier's gravitational potential energy at any point is equal to his weight multiplied by the height he has reached

Zero-gravity

Astronauts feel weightless in space because, in orbit around Earth, they are in free fall. In a circular orbit (in which a body is always at the same distance from the object it is orbiting), there would be no change in the astronaut's gravitational potential energy, since they neither move closer to nor further away from Earth.

Water bubble floating inside the International Space Station is in free fall

Chemical energy

Energy stored in chemical bonds is called chemical energy. Chemical bonds are forces that bind two or more atoms together. During a chemical reaction, atoms rearrange to form new molecules and compounds – and energy is required to break the bonds. A chemical reaction may release energy, such as when fuel burns.

Combustion reaction in flame of candle converts chemical energy in wax into heat and light

Light my fire
Fire, also known as combustion, is a reaction between oxygen in the air and the chemical energy in a fuel. It gives off heat and light. Combustion reactions help warm people's houses, power motor vehicles, and generate electricity.

Chemical energy released from a burning firework is converted to sound, light, and heat energy

Getting a reaction
When a chemical reaction occurs, one set of substances changes into another. Chemical bonds are broken and new ones are made, resulting in potential chemical energy being converted to other forms. Some chemical transformations occur slowly and may need a constant input of energy to keep them going. Others, such as fireworks, happen instantaneously, going off with a bang.

Once started, these dominos do not need further input of energy to keep them going

Hydrogen released burns and heats potassium atoms, which produce a lilac glow

Activation energy

The energy required to start a chemical reaction is called activation energy, which typically comes from heat, electricity, or light. This is similar to flicking the first domino in a chain to make all of them topple over.

Energy must be provided to make first domino fall

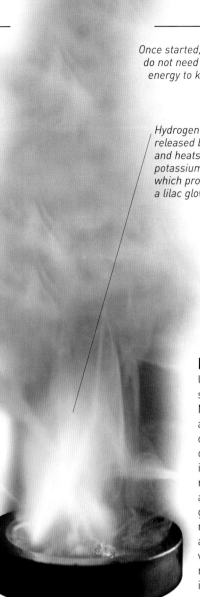

Energy transformations

Under normal conditions, most substances on Earth are stable. Most chemical reactions require an input of energy to start them off, but some reactions can give out more energy than they take in. They are called exothermic reactions (endothermic reactions absorb energy in contrast). A good example of an exothermic reaction is that between water and potassium. Water reacts vigorously with potassium to release hydrogen gas, which ignites explosively, releasing a lot of heat energy.

Harmful exhaust gases, such as carbon monoxide, enter

Catalyst breaks down toxic gases

Less harmful gases are released

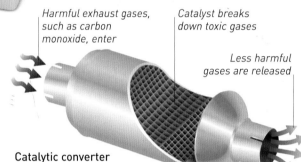

Catalytic converter

Energy enabler

Catalysts help a chemical reaction begin more easily or progress faster. They may reduce the activation energy or help a reaction create a desired result. Precious metals such as platinum or palladium are commonly used in catalytic converters in vehicles to help break down harmful exhaust gases into relatively harmless ones.

Measuring chemical energy

A calorimeter measures the change in temperature of water to determine the energy generated by a chemical reaction. Calorimeters are sealed off from the outside world. In it, a substance is burned in oxygen to increase the temperature of water. A thermometer measures the rise in temperature.

Thermometer measures temperature of water

Technician repairs tank

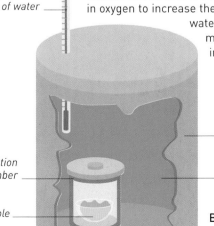

Container is sealed off from its surroundings

Water surrounds chamber

Reaction chamber

Sample

Bomb calorimeter

Energy for life

Living things are powered by chemical reactions in the cells of their bodies. Plants take radiant energy from the Sun and store it as chemical energy. Other creatures burn food to provide them with energy, but food can mean almost anything – in this aerobic digestion tank (above), bacteria feed on raw human sewage.

What is electricity?

Electricity involves charged particles – ions (atoms with extra or fewer-than-normal electrons) or the electrons themselves. A charged particle has potential energy in an electric field. The electrical energy that makes a charge move inside a conductor as a current is also easy to convert into other forms of energy.

Shocking discovery

Although the ancient Greeks noticed the strange effects of static electricity, it wasn't until much later that scientists learned how to use electrical energy as a source of power. In 1752, American Benjamin Franklin (above) sparked off the search by proposing an experiment to collect electric charge from thunderclouds.

Go with the flow

In a metal wire, current is carried by electrons, but other charged particles can do the same thing. When the rod in the middle of this sphere (below) is charged to a high voltage, atoms of gas inside the ball are pulled apart to make a plasma (a soup of ions and electrons). Electricity builds up in the centre and then zaps through the plasma, producing streams of mini lightning bolts.

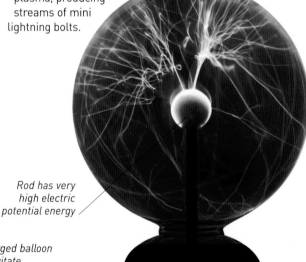

Rod has very high electric potential energy

Positively charged balloon makes hair levitate

Opposites attract

Rubbing certain materials together causes electrons to move, leaving some areas with a net positive charge and others with a net negative charge. Since opposite electric charges attract, when objects build up unequal charges, the electric potential energy can make charges move to even out the imbalance. This is called static electricity. Charge differences can build up – on a balloon rubbed against dry hair, for example.

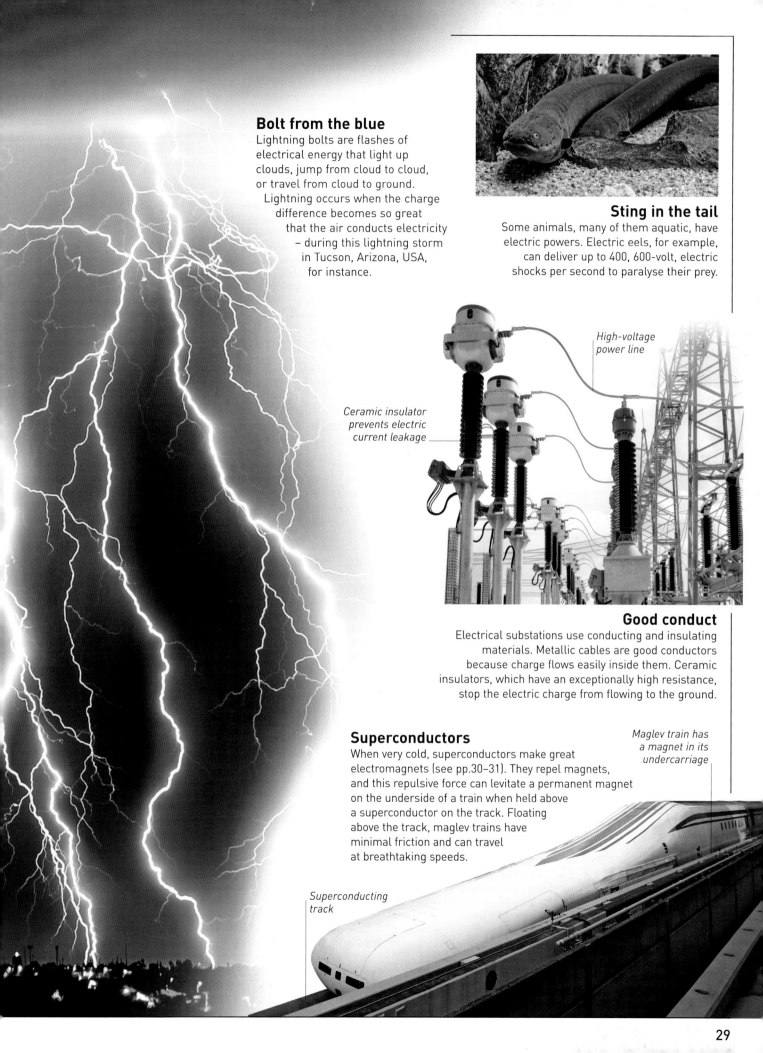

Bolt from the blue

Lightning bolts are flashes of electrical energy that light up clouds, jump from cloud to cloud, or travel from cloud to ground. Lightning occurs when the charge difference becomes so great that the air conducts electricity – during this lightning storm in Tucson, Arizona, USA, for instance.

Sting in the tail

Some animals, many of them aquatic, have electric powers. Electric eels, for example, can deliver up to 400, 600-volt, electric shocks per second to paralyse their prey.

High-voltage power line

Ceramic insulator prevents electric current leakage

Good conduct

Electrical substations use conducting and insulating materials. Metallic cables are good conductors because charge flows easily inside them. Ceramic insulators, which have an exceptionally high resistance, stop the electric charge from flowing to the ground.

Superconductors

When very cold, superconductors make great electromagnets (see pp.30–31). They repel magnets, and this repulsive force can levitate a permanent magnet on the underside of a train when held above a superconductor on the track. Floating above the track, maglev trains have minimal friction and can travel at breathtaking speeds.

Maglev train has a magnet in its undercarriage

Superconducting track

One of
two poles

Marvellous magnets

Permanent magnets are created when the "spin" of negatively charged electrons in a magnetic material are made to line up. Countless tiny magnetic fields add together to produce the forces felt between magnets and magnetic materials. These forces mean that magnetic objects have potential energy in magnetic fields. Moving magnetic materials in a magnetic field changes the object's energy.

Iron filings around a magnet reveal its magnetic field

Most filings are attracted to a pole

Spoon made of a magnetic rock, or lodestone

Polished bronze plinth

Electromagnetism

In the 19th century, scientists discovered a link between magnetism and electricity. The electromagnetic force connects electric and magnetic fields. It makes opposite electric charges (or magnetic poles) attract and like charges (or poles) repel. Attraction between electrons and protons keeps the atoms in matter from falling apart. Energy can be given to and taken from particles moving through these force fields.

Strange attractors

The ancient Chinese invented the compass between 2 BCE and 1 CE. Early versions had a spoon-shaped pointer, which aligned with Earth's magnetic field. Navigators used the north and south directions on the compass to find their way at sea.

Magnetic discovery

In 1820, Danish scientist Hans Christian Ørsted (left) discovered that electricity can generate magnetic fields. While setting up an experiment, he noticed the needle of a compass moved when a nearby coil of wire was conducting electricity. Electrical energy was changing into magnetic energy. This hinted that different forms of energy are related.

Motor effect

Motors convert electrical energy into kinetic energy. In a motor, electric charges flowing through a conducting wire coil produce a magnetic field around the wire. When placed in a magnetic field, the wire feels a force of attraction or repulsion, which makes the coil rotate. Electric motors are used to power many machines, such as a drill.

Electric motor powers drill

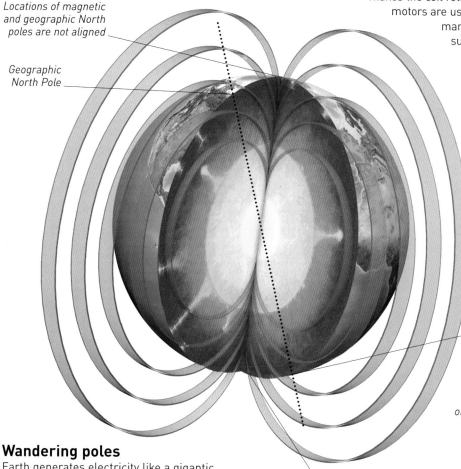

Locations of magnetic and geographic North poles are not aligned

Geographic North Pole

Earth's magnetic field

Geographic South Pole

Magnetic South pole drifts northwards by about 10–15 km (6–9 miles) per year

Powerful magnets

An electric current in a wire induces a magnetic field. If the wire is coiled around an iron core, then the electromagnet becomes much stronger. Electromagnets can be used to separate metal in places like scrap yards (such as the one below).

Giant electromagnet on crane can lift huge weights

Wandering poles

Earth generates electricity like a gigantic dynamo. Circulation within the planet's liquid metallic core sets up electric currents, which in turn generate Earth's magnetic field. The North and South magnetic poles move over time, and their locations do not always match those of the planet's geographic North and South poles.

Dynamic electricity

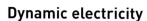

Moving a magnet inside a coil of wire produces electricity. This is the way most of our electricity is generated. A linear induction torch (left) converts a magnet's kinetic energy (created by shaking the torch) into electric energy, which an LED array converts to light energy.

Nuclear energy

Atoms consist of a nucleus orbited by electrons. The nucleus contains particles called protons and neutrons. Very strong forces hold the nucleus together, and these forces are involved in nuclear reactions. Nuclear reactions are very powerful, and result in some mass being lost and changed into energy.

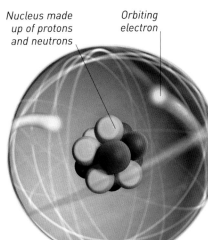

Nucleus made up of protons and neutrons

Orbiting electron

Inside an atom
The nucleus, containing positively charged protons and uncharged neutrons, is in the centre of an atom. Moving around the nucleus are negatively charged electrons. Atoms are tiny – 173 million times smaller than a penny.

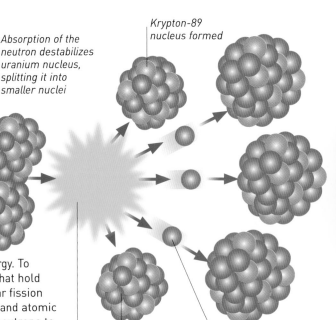

Krypton-89 nucleus formed

Absorption of the neutron destabilizes uranium nucleus, splitting it into smaller nuclei

Target nucleus (unstable uranium-235 nucleus)

Neutron

Barium-144 nucleus formed

Fast-moving free neutrons released, which can each go on to split further uranium nuclei in a chain reaction

Energy is released when atomic nucleus splits

Chain reaction
An atom's nucleus is a store of potential energy. To access that energy, the super-strong forces that hold the nucleus together must be broken. Nuclear fission splits the nucleus. In nuclear power stations and atomic bombs, each fission reaction supplies more neutrons to split ever more nuclei, causing a chain reaction.

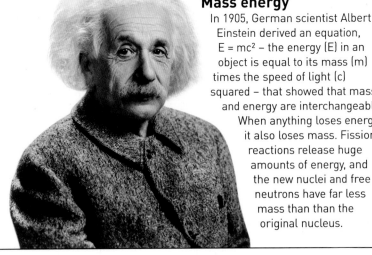

Mass energy
In 1905, German scientist Albert Einstein derived an equation, $E = mc^2$ – the energy (E) in an object is equal to its mass (m) times the speed of light (c) squared – that showed that mass and energy are interchangeable. When anything loses energy, it also loses mass. Fission reactions release huge amounts of energy, and the new nuclei and free neutrons have far less mass than than the original nucleus.

A mushroom cloud is characteristic of huge amounts of heat energy released suddenly

Atomic explosions rise significantly higher into the atmosphere than any other type of explosion

Across the Universe

Most satellites and space probes are powered by solar energy. However, as these travel further away from the Sun, this energy source gets weaker. Space probes that hope to travel past the orbit of Jupiter must have an alternative means of powering their systems. *New Horizons*, the NASA probe that travelled to Pluto in 2015, has its own mini nuclear power plant on-board.

Nuclear threat

In a nuclear reactor, control rods can soak up excess neutrons, maintaining the optimum number required to sustain a chain reaction, or they can close it off entirely. In a nuclear bomb, the chain reaction is uncontrolled. Runaway fission produces so much energy so quickly that it results in an enormous explosion. The biggest bomb ever built was Russia's 27-tonne Tsar Bomba. When this was tested in 1961, it exploded with ten times the entire firepower used in World War II.

Risky business

Nuclear energy is a very efficient method of generating electricity, and does not cause global warming. However, there is always the risk that accidents, such as at Chernobyl (above) or Fukushima, will release deadly radioactive material into the environment. Nuclear reactors are also used to produce weapons-grade material for atomic bombs.

Conservation of energy

Energy can neither be created nor destroyed. This fact holds true wherever you go in the Universe. The idea of conservation of energy says that nothing comes for free: if you want to get energy out, you need to put energy in. Energy can only change in form, and can easily be transferred from one form to another.

Computer-generated image showing particle collisions inside a particle accelerator

Forever constant

Conservation of energy means that the total energy in a closed system is constant. In a particle accelerator, such as at CERN, Geneva, for example, the kinetic energy of colliding particles is converted into other types of energy, and can even create new particles of matter. However, the total amount of energy in the accelerator always remains the same.

Law of nature

In 1841, German physician Julius Robert von Mayer (above) came up with the idea that energy is neither created nor destroyed. But it was the British scientist James Joule who got the credit for the idea a year later. Joule's experiments showed that a falling weight could heat water, proving that energy did not disappear, but was converted into different forms.

Converting energy

Things never run out of energy. Energy may be converted from one form to another, but the total amount is conserved. Chemical energy from food powers the muscles in our body that send a bowling ball racing towards the pins – providing it with lots of kinetic energy. This kinetic energy is then transferred to the pins, sending them flying. In the process, some of the energy is also converted to sound and heat.

Most of the ball's kinetic energy is transferred to the pins in a collision, but some is converted to sound energy

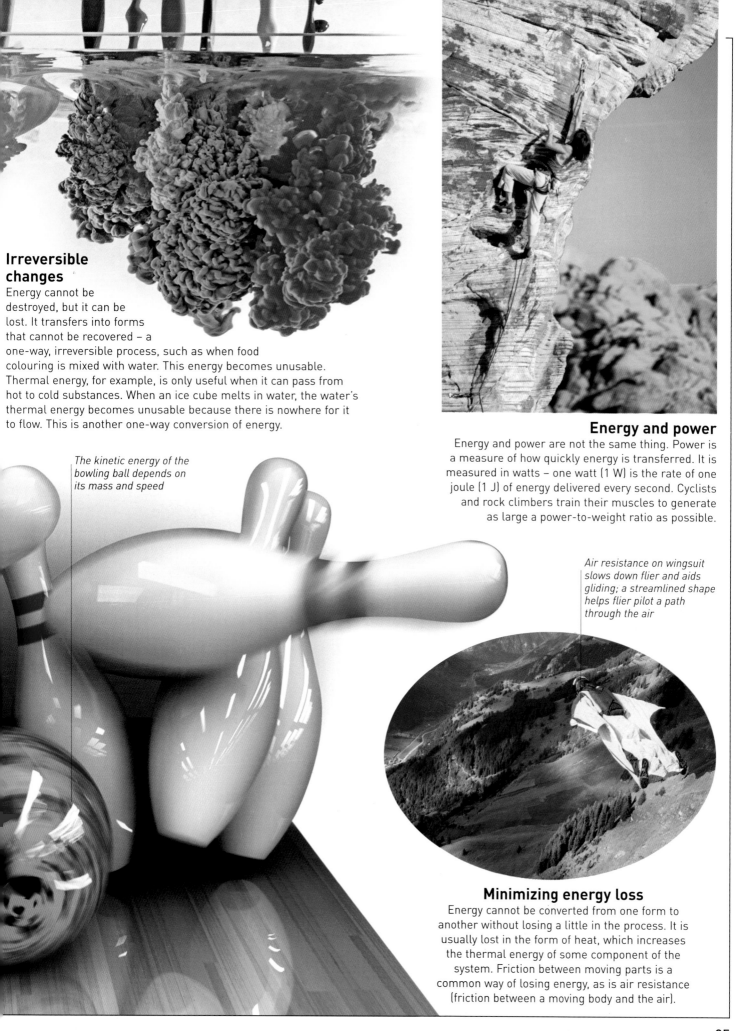

Irreversible changes

Energy cannot be destroyed, but it can be lost. It transfers into forms that cannot be recovered – a one-way, irreversible process, such as when food colouring is mixed with water. This energy becomes unusable. Thermal energy, for example, is only useful when it can pass from hot to cold substances. When an ice cube melts in water, the water's thermal energy becomes unusable because there is nowhere for it to flow. This is another one-way conversion of energy.

The kinetic energy of the bowling ball depends on its mass and speed

Energy and power

Energy and power are not the same thing. Power is a measure of how quickly energy is transferred. It is measured in watts – one watt (1 W) is the rate of one joule (1 J) of energy delivered every second. Cyclists and rock climbers train their muscles to generate as large a power-to-weight ratio as possible.

Air resistance on wingsuit slows down flier and aids gliding; a streamlined shape helps flier pilot a path through the air

Minimizing energy loss

Energy cannot be converted from one form to another without losing a little in the process. It is usually lost in the form of heat, which increases the thermal energy of some component of the system. Friction between moving parts is a common way of losing energy, as is air resistance (friction between a moving body and the air).

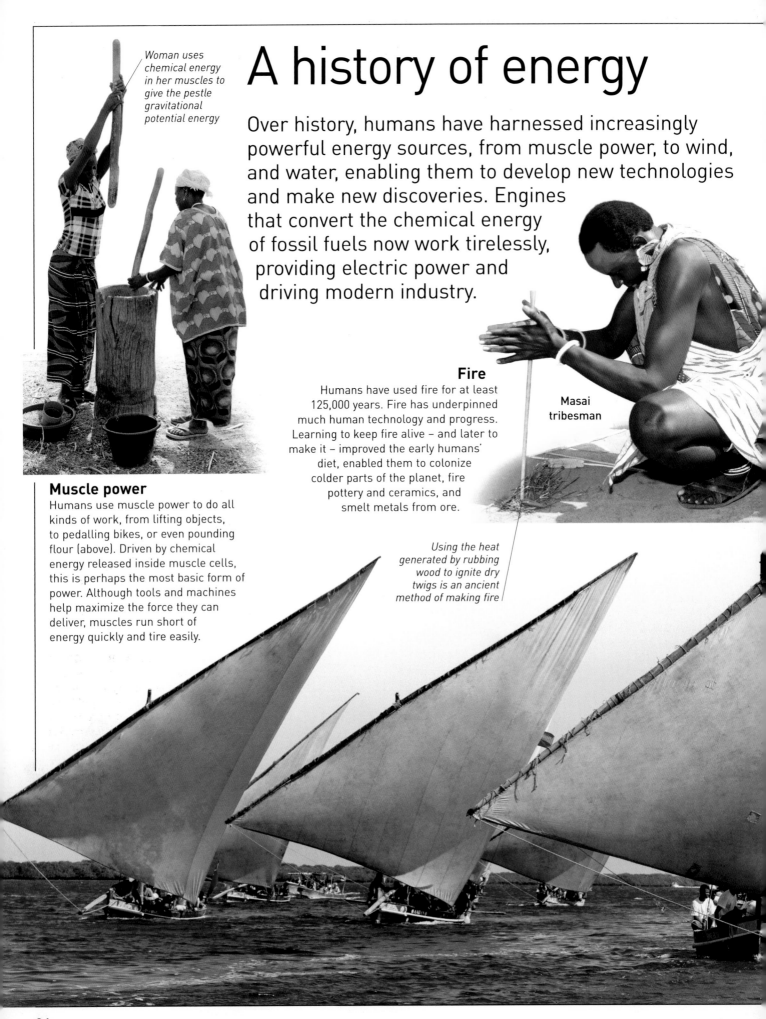

Woman uses chemical energy in her muscles to give the pestle gravitational potential energy

A history of energy

Over history, humans have harnessed increasingly powerful energy sources, from muscle power, to wind, and water, enabling them to develop new technologies and make new discoveries. Engines that convert the chemical energy of fossil fuels now work tirelessly, providing electric power and driving modern industry.

Fire

Humans have used fire for at least 125,000 years. Fire has underpinned much human technology and progress. Learning to keep fire alive – and later to make it – improved the early humans' diet, enabled them to colonize colder parts of the planet, fire pottery and ceramics, and smelt metals from ore.

Masai tribesman

Muscle power

Humans use muscle power to do all kinds of work, from lifting objects, to pedalling bikes, or even pounding flour (above). Driven by chemical energy released inside muscle cells, this is perhaps the most basic form of power. Although tools and machines help maximize the force they can deliver, muscles run short of energy quickly and tire easily.

Using the heat generated by rubbing wood to ignite dry twigs is an ancient method of making fire

Making metals

As well as releasing warmth, the energy from fire can cause chemical changes. Smelting extracts metallic elements that are chemically bound to other components in ores. This process requires high temperatures. The process of smelting allowed people to make hard, long-lasting implements out of metals such as bronze and iron.

An artwork showing humans smelting iron

Animal power

The technology of any age is linked to the amount of power available. While animal muscle can generate more energy than human muscle, and for longer, it is not enough to power vehicles such as aircraft. However, animals can do tasks that would be exhausting or impossible for one person, such as pulling ploughs.

Harnessing the wind

The boat was one of the earliest forms of powered transport. In sailing boats, cloth sails capture the natural motion of the air to move the boats across water. On land, windmills harnessed the energy of wind to power millstones, pump water, and grind grain to make bread.

Water power

Until the invention of the steam engine, water power was the main source of energy for the factories of the Industrial Revolution. Pulleys coming off a central driveshaft, powered by a water wheel (such as the one above), ran large rooms full of machines. The energy to turn the wheel came from the kinetic energy of falling water.

Dhows participating in a dhow race off Lamu Island, Kenya

Steaming ahead

The invention of the steam engine changed everything. Powered by the chemical energy of fossil fuels, it allowed a machine to work without rest. Factories powered by steam engines no longer needed to be close to water. People began to use steam engines to transport passengers and goods.

Water stored in the boiler is heated and turned into steam

Fossil fuels

Fossil fuels (coal, oil, and natural gas) are made of the carbon-rich remnants of organisms that were alive hundreds of millions of years ago. The energy they contain can be released through burning, but once consumed fossil fuels cannot be replaced.

Industrial revolution

Coal was the first fossil fuel to be to be used on a large scale. The sheer density of chemical potential energy concentrated in coal could power steam engines, which mechanized manufacturing in the 19th century.

How fossils become fuel

Oil forms when microscopic plankton living in the oceans die and accumulate on the sea bed. Buried rapidly by sediment, the organic remains are squeezed, heated, and altered chemically to become oil and natural gas. Coal forms by a similar process, but instead of marine creatures, it is formed by the remains of plants from ancient swamp forests.

Millions of microscopic plankton float in the sea

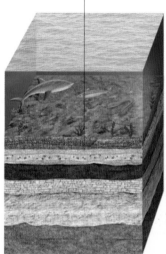
Sediment layer is buried by mud

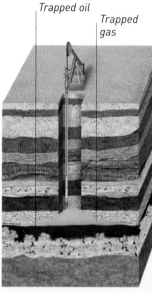
Trapped oil
Trapped gas

1 Ancient creatures
Dead plankton settle on sea bed and are slowly covered with sediment.

2 Conversion to oil
Bacteria, pressure, and heat convert plankton into oil and natural gas.

3 Drilling
Oil and natural gas form reservoirs, which humans can tap.

Refined fuel

Oil found in the ground is called crude oil. Before it can be used, it needs to be refined by a process called fractional distillation. This involves heating crude oil until it turns into vapour. Different parts of the vapour, called fractions, condense at different temperatures. The fractions are then processed to make a range of fuels.

Crude oil distillation tower

Black gold

Oil is the currency of our modern world. The entire transport system runs on it. The chemical potential energy it contains is used to generate electricity that powers factories and industries, heats our homes, and enables us to use many household appliances.

A nodding donkey, or pumping jack, lifts oil from wells

Life's a gas

Natural gas is a mix of hydrocarbons. It provides energy to cook food and heat homes. Gas is tapped off the top of oil reservoirs, or sometimes forced out of oil shale rocks by hydraulic fracturing, also known as fracking. Natural gas is odourless, but a chemical is added to it to give it a distinct smell, so that it is easy to detect in case of a leak.

Burning ice

Strange substances called methane clathrates are found under sediments on the ocean floor. These are ices that have methane gas trapped inside their structures. Surprisingly, these ices burn. The large quantities of methane gas inside them make them an important potential fuel source for the future.

Disaster zone

Our dependence on fossil fuels creates many problems. Oil spills or leaks are ecological disasters that devastate ecosystems for decades. They may even lead to a fire, as seen in the image above. Burning fossil fuels releases chemicals that cause air pollution and acid rain. Carbon dioxide released by burning fossil fuels also adds to global warming.

Exhaust pipe for fired heater that heats oil to about 350°C (660°F)

Engines

The invention of engines, such as the steam engine and the internal combustion engine, revolutionized transport. Mechanical engines need to be compact enough to be carried within the vehicle they are powering. They run constantly, converting the potential chemical energy of fuel into kinetic energy. Efficiency is the name of the game in engines: how to convert as much of the potential energy into usable power as possible.

Crankshaft
A crankshaft converts rotary motion into back-and-forth, linear motion. The illustration above shows a drawing of the crankshaft invented by medieval Islamic engineer al-Jazari in 1206. The crankshaft in an internal combustion engine converts up-down motion into the rotary motion that powers a vehicle.

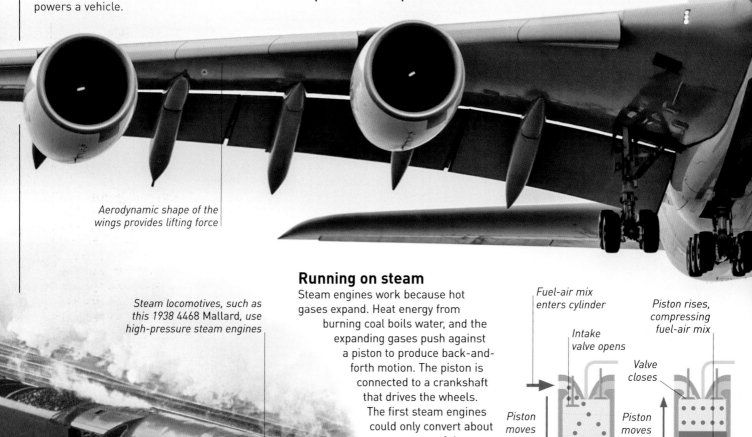

Aerodynamic shape of the wings provides lifting force

Steam locomotives, such as this 1938 4468 Mallard, use high-pressure steam engines

Running on steam
Steam engines work because hot gases expand. Heat energy from burning coal boils water, and the expanding gases push against a piston to produce back-and-forth motion. The piston is connected to a crankshaft that drives the wheels. The first steam engines could only convert about one per cent of the available energy.

Fuel-air mix enters cylinder

Intake valve opens

Piston moves down

Fuel has chemical energy

1. Intake

Piston falls, pulling in fuel-air mix

Piston rises, compressing fuel-air mix

Valve closes

Piston moves up

2. Compression

Jet power

Huge aeroplanes, such as the Airbus A380 (below), need powerful engines to get them off the ground. Most use turbofan engines (a kind of jet engine). A turbofan engine sucks in and compresses air, forcing it into the combustion chamber at high pressure. Exhaust gases blasting out of the chamber turn a set of fins, which plumb the energy back to the turbine blades at the front.

Petrol-powered engine

Running engine recharges battery

High-tech hybrid

Hybrid cars use two types of power source: an internal combustion engine (that runs on fuel), and an electric motor (that turns electric energy into kinetic energy). Electric motors are much more energy efficient than internal combustion engines, converting about 80 per cent of the available energy. Vehicles using this technology help to cut down harmful emissions.

Turbofan engine burns fuel at an incredible rate, converting chemical energy to kinetic energy very quickly

Rocket engines

In a typical rocket engine, hydrogen gas in liquid form is combined with liquid oxygen in an explosive chemical reaction that sends hot gases streaming out of the rocket nozzle. The kinetic energy of the expanding gases provides enough thrust to lift a rocket against the force of Earth's gravity into space.

Liquid oxygen stored in pressurized tank

Fuel tank carries liquid hydrogen

Flow of fuel to the combustion chamber

Spark plug ignites fuel-air mix. The explosion forces piston downwards.

Exhaust valve opens, allowing burnt fuel out of cylinder

Hot exhaust gases waste energy

Flow of oxygen to the combustion chamber is controlled by a pump

Fuel and oxygen burn in the combustion chamber

Internal combustion engine

The internal combustion engine was invented in the 19th century. The engine ignites fuel inside a metallic cylinder, called a piston. The energy of the exploding fuel sends the piston whizzing up and down as much as 6,000 times per minute. This system only manages to use about 24 per cent of the energy generated.

Nozzle

3. Power

4. Exhaust

Crankshaft turns up-and-down motion of piston into rotary motion of wheel in a car

Piston rises, forcing out exhaust gases

Hot exhaust gases

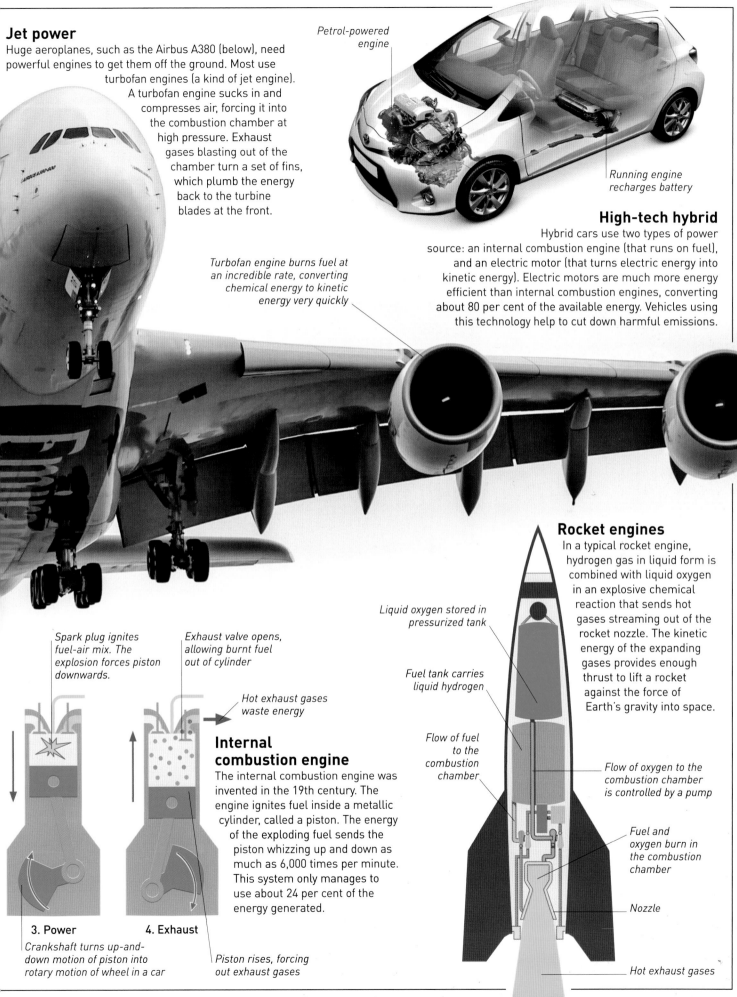

Generating electricity

Power stations turn the chemical energy in fossil fuels into electrical energy that can be used in homes and businesses. Electricity is generated by heating water to make high-pressure steam, which drives the turbines and generators.

Key

- Space cooling
- Lighting
- Water heating
- Space heating
- Refrigeration
- Running household appliances

Power in the home

The average US household uses around 20,000 kWh of energy every year. That is the same amount of chemical potential energy released by burning 12 barrels of oil. Most of this energy is used for running household appliances and for keeping homes at a comfortable temperature.

Power struggle

The "War of the Currents" was a battle in the late 1880s over what system was better for transmitting electrical energy: direct current (DC) – in which electricity flows in one direction – and alternating current (AC), in which the direction of electric current reverses several times a second. AC, promoted by Nikola Tesla (right), won the day, but DC is still used inside most electronic devices, and must be converted back from AC.

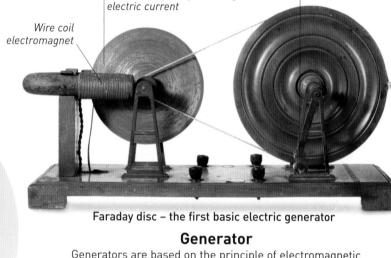

Copper disc spins between poles of magnet, generating electric current

Turning crank

Wire coil electromagnet

Faraday disc – the first basic electric generator

Generator

Generators are based on the principle of electromagnetic induction, discovered in 1831 by British scientist Michael Faraday. According to this principle, an electromagnetic force is produced when an electrical conductor moves across a magnetic field.

Smoke stacks from the furnaces are tall so that exhaust gases can disperse easily

Furnace

Dirty fuels

Burning fossil fuels releases carbon dioxide into the atmosphere, which contributes to global warming. Combustion also releases sulphur dioxide, ozone, nitrous oxides, and tiny dust particles (called particulates), which cause smog in many cities where they are allowed to build up, such as Shanghai (above).

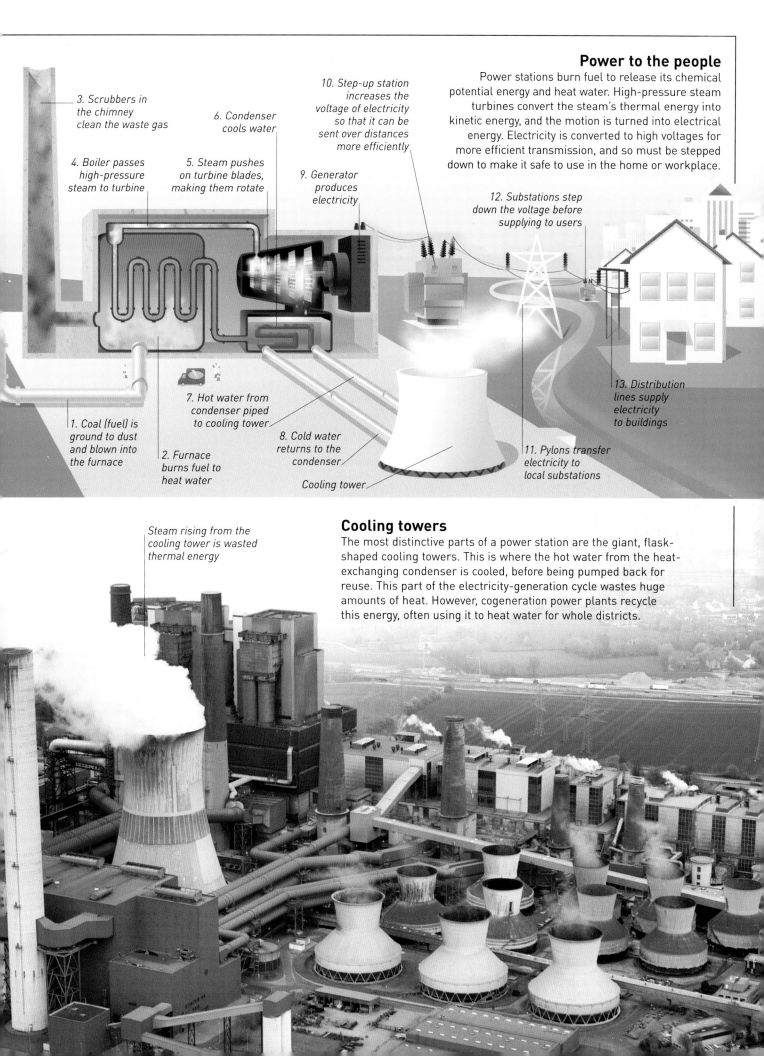

Power to the people

Power stations burn fuel to release its chemical potential energy and heat water. High-pressure steam turbines convert the steam's thermal energy into kinetic energy, and the motion is turned into electrical energy. Electricity is converted to high voltages for more efficient transmission, and so must be stepped down to make it safe to use in the home or workplace.

3. Scrubbers in the chimney clean the waste gas

6. Condenser cools water

10. Step-up station increases the voltage of electricity so that it can be sent over distances more efficiently

4. Boiler passes high-pressure steam to turbine

5. Steam pushes on turbine blades, making them rotate

9. Generator produces electricity

12. Substations step down the voltage before supplying to users

7. Hot water from condenser piped to cooling tower

1. Coal (fuel) is ground to dust and blown into the furnace

8. Cold water returns to the condenser

2. Furnace burns fuel to heat water

Cooling tower

11. Pylons transfer electricity to local substations

13. Distribution lines supply electricity to buildings

Cooling towers

The most distinctive parts of a power station are the giant, flask-shaped cooling towers. This is where the hot water from the heat-exchanging condenser is cooled, before being pumped back for reuse. This part of the electricity-generation cycle wastes huge amounts of heat. However, cogeneration power plants recycle this energy, often using it to heat water for whole districts.

Steam rising from the cooling tower is wasted thermal energy

Using energy in the home

We use a lot of energy in the home: to keep it warm in winter and cool in summer; to freeze food, to keep it cool, and even to cook it. Adequate insulation stops or reduces heat from escaping, and helps make homes much more energy efficient.

Heat pumps

Refrigerators work by moving energy from one place to another. A coolant is compressed to become liquid and is then piped into the cold box. Inside, it expands, drawing thermal energy away from the air in the box as it turns back into a gas. Pumped back outside, the coolant is compressed again. A metal grid helps the liquid lose its heat.

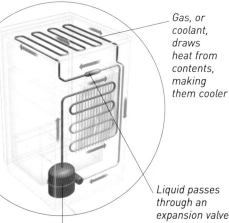

Gas, or coolant, draws heat from contents, making them cooler

Liquid passes through an expansion valve where it turns into gas

Compressor turns gas back into liquid

Induction heating

An electromagnet in the induction hob produces a constantly fluctuating electric current in a coil. The changing magnetic field induced by a metal saucepan heats it and the food inside it. It can also be used in dynamos (electrical generators) to produce electrical energy from the kinetic energy of moving magnets.

Heating

Keeping the house warm or cool, and heating water requires energy. The chemical potential energy of natural gas, wood, or oil can also be released by burning it in stoves, heaters, or boilers. Combustion reactions in a wood-burning stove (right) heat the surrounding air. This heated air then circulates around the room and warms it up via convection.

Cooking with waves

A microwave oven uses high-energy radio waves to cook food. Water molecules in food absorb the energy, and the increased thermal energy heats and cooks the food. Electromagnetic radiation also communicates with an Internet wireless router using radio waves.

Gadgets and gizmos

Most toys and games need energy to function. Some use elastic potential energy, such as clockwork toys, but many use electrical energy. The motors driving the rotors of the drone below convert electrical energy (transferred from the chemical energy of batteries) into kinetic energy.

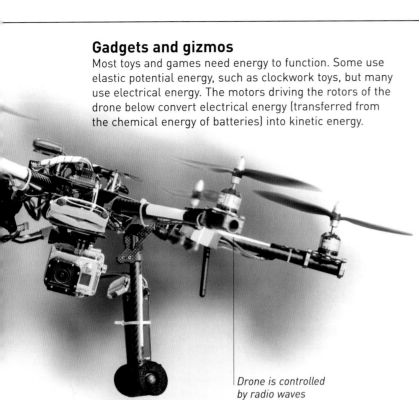

Drone is controlled by radio waves

Saving energy

Without insulation, heat energy escapes easily through a house's roof, walls, windows, and doors. Seen above is an imaging scanner that can detect temperature variations in a house – the red areas indicate portions with greater heat loss. Installing insulation under the roof, and in the gap between the inner and outer walls, reduces these energy losses and makes homes more energy efficient.

Eco housing

The zero carbon-emission house shown here is built using materials that are energy efficient. The large triple-glazed windows trap radiant energy, while high thermal-mass walls store the energy from the Sun. This results in ultra-low energy usage. Solar panels on the roof supply electricity and hot water, and can even provide more energy than the house needs.

Solar panel converts radiant energy to electrical energy

Storing energy

Energy cannot be created or destroyed, but it can be stored. Storing energy transfers it into potential energy that can be used later. Nature is good at doing this: plants store the Sun's energy as glucose, while animals store energy as fat. Energy must be expended to store energy.

Flash capacitors

Capacitors store electric charge on a pair of plates, separated by a minuscule gap. In electronic devices, capacitors smooth out current, provide backup power for memory, and release sudden jolts of energy when needed – to power a camera flash, for example.

Piles of electricity

Italian scientist Alessandro Volta invented the first battery in 1800. Called the Voltaic pile, this battery (left) was composed of a stack of copper and zinc discs washed in acid. The acid and metals interacted in a series of chemical reactions, creating a flow of electrons to generate a continuous electric current.

Copper disc

Zinc disc

Voltaic pile

Portable power

Rechargeable batteries, such as lithium-polymer mobile phone batteries and lead-acid car batteries, can be restocked with energy to use later. Although, in theory, rechargeable batteries should be reusable forever, their ability to store energy decreases over time.

Portable charging battery and media player

Energy in motion

A flywheel is a spinning disc that stores kinetic energy. The motion energy of a vehicle, such as a locomotive, can be transferred to a flywheel to slow it down, and kept ready to be fed back to provide a speed boost when needed. This is also how the mechanical kinetic energy recovery system (KERS) used in F1 motorcars works.

Flywheel system fitted at the axle recovers much of the car's kinetic energy as it slows down

Magic instant heat

"Phase-change" materials transform their state from solid to liquid and back again when exposed to certain temperatures. They release energy as they turn solid and absorb energy when they become liquid. In a sodium acetate heat pad, energy from a metal "clicker" causes the liquid to crystallize, releasing energy as heat when it turns solid.

Heat pads made out of phase-change materials can be used to relieve pain

Exhausted batteries can leak harmful chemicals

Fuel cells

Using electrical energy to split water into hydrogen and oxygen is another way of storing energy. These two reactive gases can later recombine to release their energy in a chemical reaction. Fuel cells are devices for converting chemical energy into electrical energy. Unlike a battery, they keep generating power so long as they receive fuel.

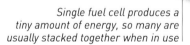

Single fuel cell produces a tiny amount of energy, so many are usually stacked together when in use

Environmental cost of batteries

Most batteries contain so-called "heavy metals", such as mercury, lead, cadmium, and nickel, which are all harmful to the environment. As well as poisoning wildlife, heavy metals can cause serious, life-threatening illnesses when they pollute water supplies, and tend to build up in the food chain.

Engine (housed in this part of the car) can benefit from power boosts provided by the KERS

Energy for communication

Sending messages from one place to another requires energy. The signals that carry information – electricity, light waves, radio waves, and sound waves – are all forms of energy. They must be powerful enough to travel over large distances; if not, the information becomes garbled and is lost.

Losing energy

All communication systems need energy to work and are subject to loss of energy. This loss is called attenuation. For example, your voice will carry over a short distance, but if you use a megaphone, your voice will carry over a greater distance. Sending messages via electrical signals or radio waves can help reduce this loss.

Give me a buzz

The advent of electricity made communicating easier, with messages zipping along wires as pulses of electrical energy. In 1833, German inventors Carl Friedrich Gauss and Wilhelm Weber built an early telegraph system in Göttingen, linking a distance of 1 km (0.6 miles). Around 1835, American inventor Samuel Morse invented the Morse code to transmit messages. In it, letters of the alphabet are represented by short and long electrical pulses, called dots and dashes.

Dots and dashes of Morse code

Satellite communication

Communications satellites enable phone calls, TV, radio, video, and Internet traffic to be beamed anywhere in the world by receiving and sending energy in the form of radio waves. To get the signals into space and back with minimal loss of information requires high-energy microwaves. The first practical communications satellite to be launched was *Telstar 1* (left), in 1962.

Antenna

Solar cells tapped energy from the Sun

Wireless days

Wireless telegraphy messaging made ship-to-ship and ship-to-shore communication possible. Messages were encoded using Morse code, sent as blips of electrical energy that produced radio wave pulses. This radiant energy travelled through the air, and was picked up by a receiving antenna, where it was converted back to electrical energy, and finally transmitted as sound in an operator's earpiece.

Phone mast disguised as palm tree

Communication on the move

Mobile phones work by using low-intensity microwaves – a form of radiant energy. This electromagnetic radiation is absorbed by the air and loses energy easily. To combat this, a network of mobile masts relays these microwaves. They ensure that calls can be made without losing signal as a user moves around.

Interstellar communications

As well as darting around the planet, our TV and radio transmissions also travel out into space. The very first television signals will now be arriving at star systems about 80 light years away, but will be too weak to make out. Radio signals beamed out to certain zones of the galaxy from radio telescopes have more focussed energy and will travel much further. Arrays of telescopes, such as the Allen Telescope Array, in California, USA (above), search for radiant energy from space to look for signs of intelligent alien civilizations.

Communication over videos is generally "live"

Connecting the world

Having a face-to-face conversation with anyone in the world over the Internet is one of the easiest ways of communicating today. Data transmitted on the Internet runs on servers, which need to be kept running 24 hours a day. On the web, data is sent as energy via heavy-duty copper cables.

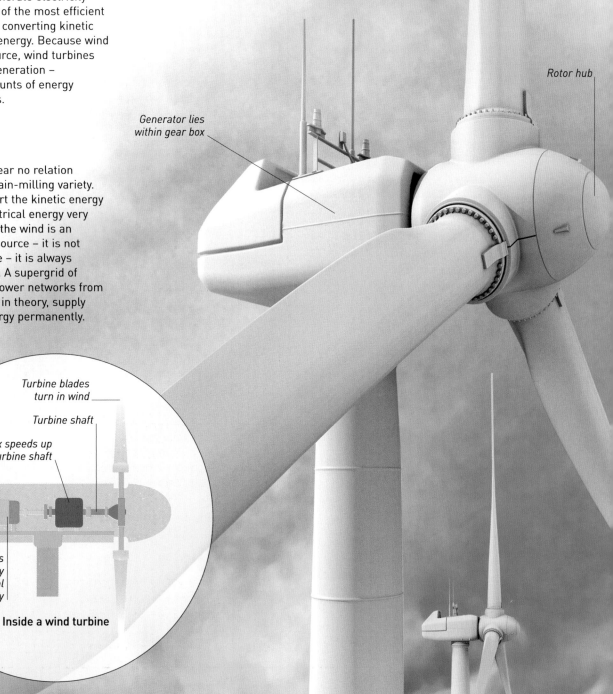

Wind energy

Wind farms are now a prominent feature of our open spaces. Unlike fossil fuels, wind is an unlimited source of free kinetic energy. It has been harnessed since ancient times to power machines and transport. Today, wind energy is mostly used to generate electricity.

Small wind turbine fulfils some of a household's energy needs

Larger blade sweeps a wider area, generating more power

Rotor hub

Generator lies within gear box

Going off-grid

The turbines that generate electricity from wind are some of the most efficient devices available for converting kinetic energy to electrical energy. Because wind is a renewable resource, wind turbines are ideal for microgeneration – providing small amounts of energy for individual houses.

Wind turbines

Modern windmills bear no relation to the traditional, grain-milling variety. Wind turbines convert the kinetic energy of the wind into electrical energy very efficiently. Although the wind is an intermittent power source – it is not available all the time – it is always blowing somewhere. A supergrid of connected electric power networks from many nations could, in theory, supply wind-generated energy permanently.

Turbine blades turn in wind

Turbine shaft

Gear box speeds up rotation of turbine shaft

Generator converts rotational energy into electrical energy

Inside a wind turbine

Moving air

Wind is the movement of masses of air from place to place. Ultimately, its kinetic energy comes from the Sun. Unequal heating of the atmosphere results in differences in air pressure and air moves to equalize them. People have used wind power for centuries to grind grain, pump water, and to move sailboats over the seas. Wind energy can also be harnessed to enjoy sports such as windsurfing.

Music in the air

Wind instruments use the kinetic energy of moving air to generate sounds. An Aeolian harp is a musical instrument whose strings are played by the wind. Wind causes its strings to vibrate, generating sound. This Aeolian sculpture in England works on the same principle.

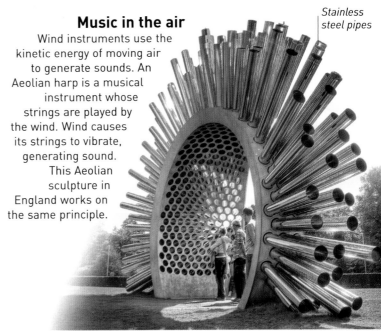

Stainless steel pipes

High-flying energy

Wind blows harder and faster in the upper atmosphere. Crosswind kite-power systems are structures that attempt to harness the energy that blows at high altitude. Shown here is the Buoyant Airborne Turbine (BAT), created to power homes in remote areas of Alaska. It has been designed to capture wind energy at great heights.

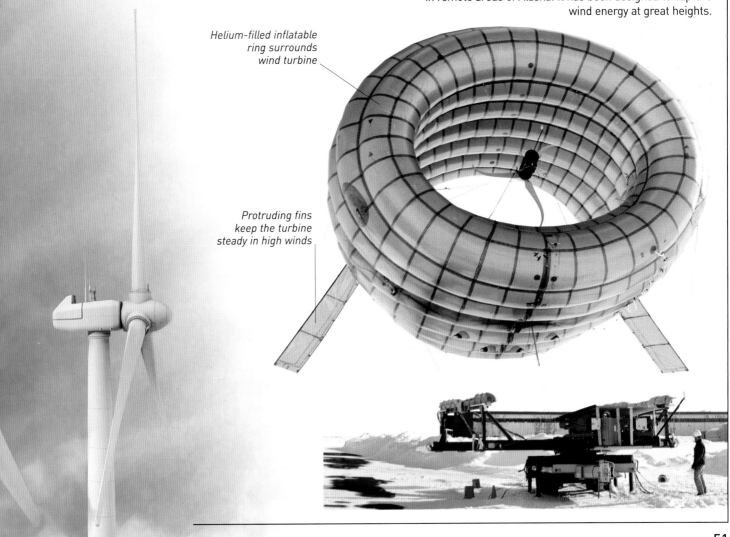

Helium-filled inflatable ring surrounds wind turbine

Protruding fins keep the turbine steady in high winds

Solar energy

Earth receives more energy from the Sun in one hour than humans use in an entire year. So far, however, we have not been able to harvest more than a tiny fraction of this energy. The Sun's energy we do harvest is called solar energy.

Sunlight cells

Solar panels are made of individual photovoltaic (PV) cells wired together. These convert solar energy into direct current (DC) electricity. Sunlight falling on a PV cell produces a tiny amount of electric potential, which drives a flow of charge.

Solar heating

On average, every square metre of Earth's surface receives several hundred joules of energy every second. Heat energy from the Sun can be used for cooking, as in the case of this solar oven in Tibet (above). It can also be used to heat water in the home.

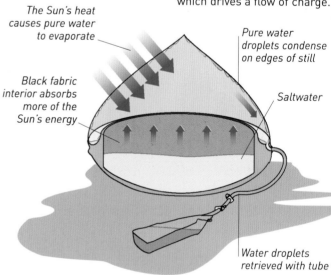

The Sun's heat causes pure water to evaporate

Pure water droplets condense on edges of still

Black fabric interior absorbs more of the Sun's energy

Saltwater

Water droplets retrieved with tube

Fresh water from the sun

Solar energy can be used to turn saltwater into fresh drinking water, using a device called a solar still. In it, saltwater is placed at the bottom of the plastic container. The water evaporates in sunlight – leaving the salt behind – and condenses at the top. The salt-free water is then collected as it slides down the sides using a tube.

Solar-powered transport

Widespread use of solar transport on the ground is still a future dream. Solar-powered vehicles include tricycles, cars, boats, and buses, and all run without a drop of fuel. In 2015, *Solar Impulse* – a Swiss aircraft that was powered entirely by light energy – was the first solar-powered aircraft to attempt a round-the-world flight.

Solar cells in the wing power four electric motors and charge lithium batteries during the day. The batteries power the aircraft at night.

Single-seater cockpit

HB-SIB

Solar furnaces

A solar furnace is a power plant for collecting the energy of sunbeams. Moveable mirrors, called heliostats, track the Sun as it moves across the sky, focussing its rays onto a central tower. The PS20, near Seville, Spain, one of the world's most powerful solar power plants, uses solar energy to boil water. This drives steam turbines to generate enough electricity to power about 8,000 homes.

Rotating mirrors reflect sunlight to a collector on top of the central tower

Solar death ray

A lens or a mirror can focus sunlight, and this concentrated beam of energy can start a fire. During the Second Punic War (218–202 BCE), the great Greek scientist Archimedes is alleged to have set attacking Roman ships on fire using mirrors to concentrate solar energy.

Power of water

After rain falls in the mountains, it flows downhill to the lowlands. As it does so, it converts gravitational potential energy into kinetic energy. Water's mass in motion can be converted into electrical energy by hydroelectric turbines. Hydropower is an inexhaustible energy supply.

Planet-shaper

A drop of rain falling from the sky converts its potential energy into kinetic energy. Over time, the energy of countless raindrops can dissolve rocks. As water flows into rivers, its mass – moving under gravity – carves out vast valleys and canyons, such as the Grand Canyon in the USA (above).

Hydroelectricity

A hydroelectric dam traps water behind a barrier to harness gravitational potential energy. When released, the water's mass transfers its potential energy to kinetic energy as it rushes downhill to a turbine. The turbine starts to rotate, and converts its energy of motion into electrical energy. The Three Gorges Dam in China (below) is the world's largest hydroelectric dam.

What a wheel

The energy of moving water has been used for centuries. Water wheels powered medieval mills that ground flour for bread, mashed wood for paper, and even hammered iron. Later, they were used to power factory machines. One type of water wheel is the noria, invented in Persia (modern-day Iran), which lifts water to a higher level to irrigate fields. Seen above is a noria in the city of Hama, Syria. With a diameter of about 20 m (65 ft), it is the largest noria in the world.

Power underground

At certain places, the planet's thermal energy, known as geothermal energy, is so close to the surface that it can be extracted and used to make electricity or heat. Cool water piped down into the ground heats up, and the steam produced can be used to drive a turbine to generate electrical energy. The Svartsengi power station in Iceland (above) produces about 475 litres of near-boiling water every second.

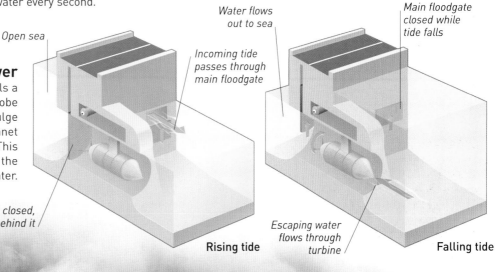

Energy from waves

The ocean is never still. The wind catches a ripple, and the friction between the air and the water pushes a wave along. Over a distance, the kinetic energy of the wind is transferred to the water. Floating machines, such as the Pelamis wave energy converter (above), can flex and bend to extract energy from the waves' ceaseless movement.

Tidal power

As it orbits Earth, the Moon's gravity hauls a huge bulge of ocean water around the globe with it. As Earth rotates, another bulge forms on the opposite side of the planet where the Moon's gravity is weakest. This causes tides. Tidal power harnesses the energy of this rise and fall of water.

Open sea

Turbine floodgate closed, trapping water behind it

Water flows out to sea

Incoming tide passes through main floodgate

Rising tide

Main floodgate closed while tide falls

Escaping water flows through turbine

Falling tide

Biopower

Biofuels, such as plants and human waste, do not contain the same quantities of chemical energy as that found in fossil fuels, but they have one crucial advantage – they are renewable sources of energy.

Burning biomass

A woman in a village in India cooks food on a stove powered by burning leftover rice stalks as biofuel. Since our planet is filled with carbon-based life forms, almost every thing that was once alive can be burned as fuel. Once an organism is dead, the organic matter in its body is called biomass.

Biofuel

Biofuels are a sustainable source of energy, made from crops that contain sugar or starch. Biofuel crops include sugarcane, sugar beet, corn, rye, sweet sorghum, and even palm oil fruit, which is harvested widely in many countries, including Malaysia (below).

Fuel up!

One advantage of biofuels is that existing petrol stations do not need to be refitted to pump this new fuel. Some gas stations in the USA provide biodiesel (made of vegetable oil). Brazilian petrol stations also serve up an alcohol called bioethanol, which is mixed with gasoline.

Green jet fuel

To power a jet plane requires a fuel that contains lots of energy. Some algae produce far more fats and oils than land plants, and can be used to create biofuel. They can be grown in both natural and artificial pools. The algae are then processed to make aviation biofuels.

Cogeneration

One power plant can be used to produce two types of energy. This method is known as cogeneration. This plant (below) uses sewage to produce biogas, which powers the generation of heat and electricity.

Kelp (a type of algae) raised on an aquaculture farm off Vancouver Island, Canada

Running on waste

Many vehicles are now driven by "green" fuels. This British bus (left) runs on bioethanol, which reduces greenhouse gas emissons and is made from renewable sources. Buses like this one help improve the sustainability of cleaner transport systems.

Fuel in fries

Waste vegetable oil from restaurants, called "yellow grease", can be converted into biofuel. It can be used to power ordinary car and truck engines. However, recovered biofuels are not carbon neutral – the process that turns yellow grease into diesel requires energy, and that energy normally comes from burning fossil fuels.

Saving energy

Most of the power we generate comes from fossil fuels. These fuels cannot be renewed. Cutting down our energy usage will reduce not only the amount of fuel we burn, but also the harmful emissions from burning fuel that end up in our atmosphere.

CFL bulb

Smart lights
Devices that convert energy more efficiently reduce the amount of fuel they use. For instance, CFLs (compact fluorescent lamps), such as the one shown above, use around one-fifth to one-third of the power of an old-style incandescent light bulb. If less fuel is burned to achieve the same effect, a smaller amount of fossil fuel needs to be extracted from the ground.

An unequal world
Not everyone has equal access to power. People in the developed world use much more energy than those in the developing world. This huge difference in energy consumption across the globe is clearly shown in this composite world map (above) made up of satellite images taken at night.

Solar panels on the roof provide electricity for the house

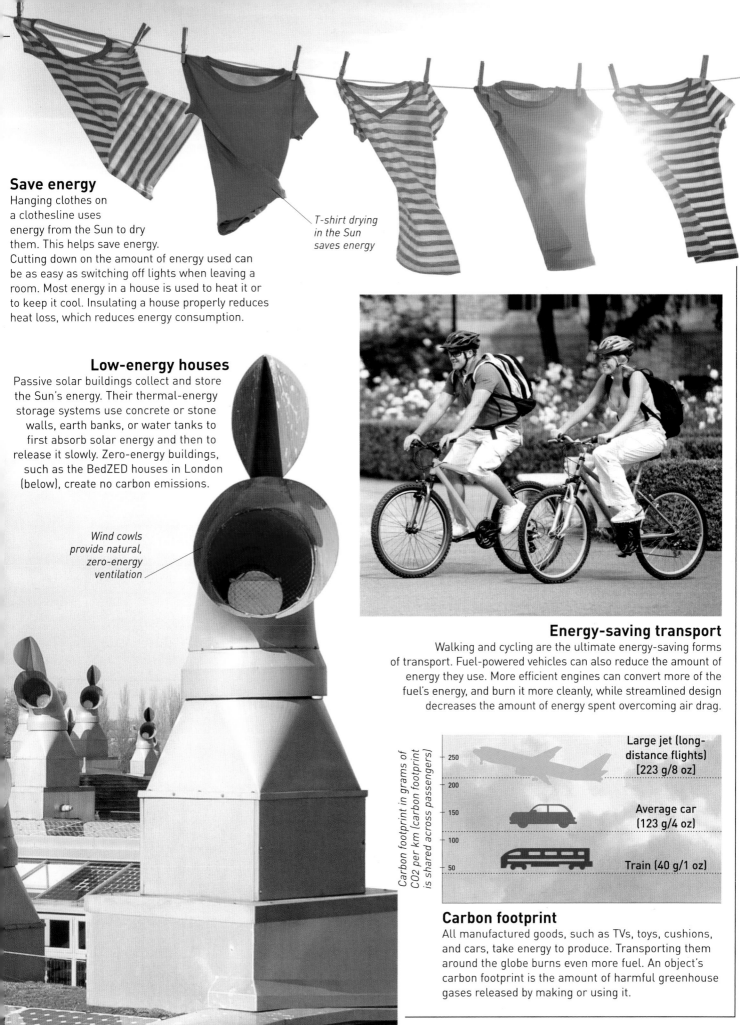

Save energy

Hanging clothes on a clothesline uses energy from the Sun to dry them. This helps save energy. Cutting down on the amount of energy used can be as easy as switching off lights when leaving a room. Most energy in a house is used to heat it or to keep it cool. Insulating a house properly reduces heat loss, which reduces energy consumption.

T-shirt drying in the Sun saves energy

Low-energy houses

Passive solar buildings collect and store the Sun's energy. Their thermal-energy storage systems use concrete or stone walls, earth banks, or water tanks to first absorb solar energy and then to release it slowly. Zero-energy buildings, such as the BedZED houses in London (below), create no carbon emissions.

Wind cowls provide natural, zero-energy ventilation

Energy-saving transport

Walking and cycling are the ultimate energy-saving forms of transport. Fuel-powered vehicles can also reduce the amount of energy they use. More efficient engines can convert more of the fuel's energy, and burn it more cleanly, while streamlined design decreases the amount of energy spent overcoming air drag.

Carbon footprint in grams of CO_2 per km (carbon footprint is shared across passengers)

250 —
200 —
150 —
100 —
50 —

Large jet (long-distance flights) [223 g/8 oz]

Average car (123 g/4 oz)

Train (40 g/1 oz)

Carbon footprint

All manufactured goods, such as TVs, toys, cushions, and cars, take energy to produce. Transporting them around the globe burns even more fuel. An object's carbon footprint is the amount of harmful greenhouse gases released by making or using it.

Future of energy

Futuristic technologies look to new ways of harnessing the forces of nature, or of developing devices to harvest small amounts of energy. Meanwhile, smart electricity grids have more control over power and demand, and switching to cleaner fuels will reduce global warming.

Virtual power grid
The power grid of the future is light on fuel and manages periods of high energy demand. A green energy grid on the Isle of Eigg, Scotland, points to the way forward. Its three hydro schemes, four wind turbines, and two solar arrays have replaced noisy and polluting diesel generators.

Flashy threads
Smart e-skin fabrics, such the one shown here, can mimic human skin – sensing pressure and temperature, stretching, and even healing itself. Future clothing may be able to generate enough electricity to power an MP3 player or to recharge a phone while you wear it.

Harnessing micropower
Instead of looking to generate large amounts of energy, smart devices and wearable technology scavenge tiny amounts of energy from body heat and motion. Piezoelectric nanogenerator floor tiles in Tokyo, Japan, harvest kinetic energy from the footfall of pedestrians.

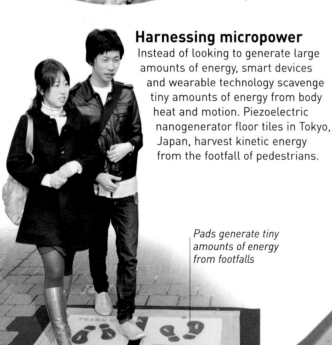

Pads generate tiny amounts of energy from footfalls

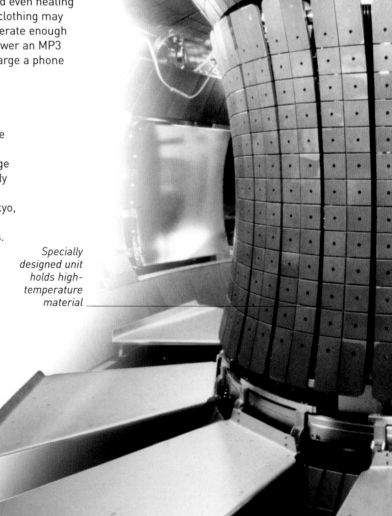

Specially designed unit holds high-temperature material

Clean dream

Hydrogen could replace fossil fuels without needing to make great changes to vehicles or filling stations. This explosive gas contains more chemical potential energy per kilogram than petrol, and takes up less space. It can be burned or combined with oxygen in a fuel cell, producing only water and no harmful gases. A car powered by hydrogen, such as the BMW H2R (left), would produce no carbon dioxide.

Airship in flight

Bio-hydrogen balloons

This futuristic concept imagines self-sufficient airships powered by hydrogen. The hydrogen is produced by a farm of seaweed. As the algae take their energy from the Sun, and hydrogen can be burned without emitting carbon dioxide, the technology is nonpolluting.

Fusion power

Nuclear fusion is a reaction that combines the nuclei of two or more atoms. Fusion reactions release enormous amounts of energy. Controlled fusion power stations would use nuclear energy to drive steam-turbine generators. However, making the reaction occur, and containing the hot material it creates takes more energy than the fusion reaction generates itself.

Power-tapping grid surrounds star

Dyson sphere

Proposed by American physicist Freeman Dyson, the Dyson sphere is a massive megastructure that surrounds a star – such as the Sun – on all sides and harnesses 100 per cent of its energy. The sphere could be solid or made of a swarm of satellite solar panels.

Universe of energy

Stars and galaxies release radiant energy, while the gravitational potential energy of all objects in space gets converted to kinetic energy. Space may not be a vacuum; it might be full of "dark energy".

Something from nothing

Every galaxy in the Universe contains thousands of stars, each of which is generating huge amounts of energy. However, the law of conservation of energy says that energy cannot be created or destroyed, so where did the energy and matter come from? The answer may be that the Universe has zero energy: matter has energy, but gravity has negative energy, and the two cancel each other out.

Expanding universe

Whichever direction you look in the sky, all the distant galaxies are moving further away. This means that galaxies were once closer to each other. Our Universe must have had a beginning, when all the galaxies were on top of each other. Our Universe is now expanding in all directions.

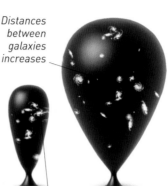

Galaxies continue to move apart from one another

Distances between galaxies increases

Galaxies were closer together in the past

Photons of light energy stop being absorbed. Universe becomes visible.

The first stars form 560 million years after the Big Bang

Big Bang

Atoms of hydrogen and helium form around 380,000 years ago

The Big Bang

Around 13.7 billion years ago, an enormous blast of energy created the Universe. Immediately before this "Big Bang", the whole Universe was contained within an extremely tiny space that had mind-blowing amounts of potential energy. In the next fraction of a second, the Universe ballooned in size, powered by dark energy.

Horn antenna at Bell Labs, New Jersey, USA

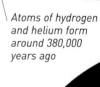

First light

If the Big Bang theory has any truth, observations must agree with its predictions. In 1964, American astrophysicists Arno Penzias and Robert Wilson (left) found a faint microwave signal coming from every direction in the sky. This was the 13-billion-year-old afterglow of the energy released during the Big Bang.

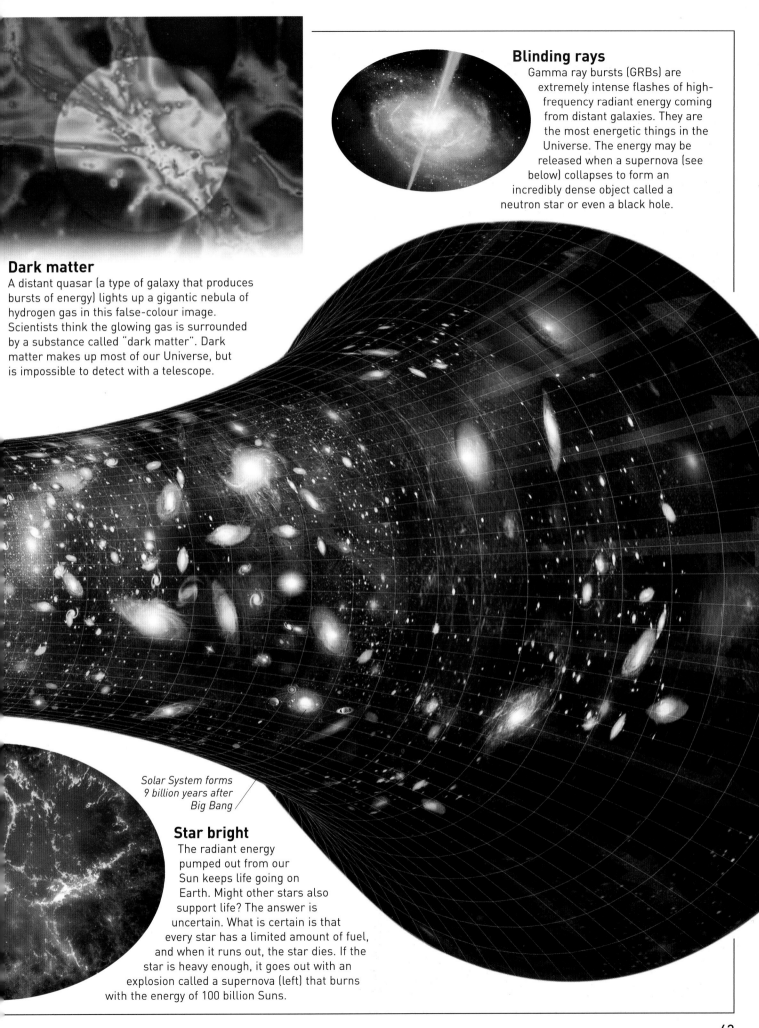

Blinding rays
Gamma ray bursts (GRBs) are extremely intense flashes of high-frequency radiant energy coming from distant galaxies. They are the most energetic things in the Universe. The energy may be released when a supernova (see below) collapses to form an incredibly dense object called a neutron star or even a black hole.

Dark matter
A distant quasar (a type of galaxy that produces bursts of energy) lights up a gigantic nebula of hydrogen gas in this false-colour image. Scientists think the glowing gas is surrounded by a substance called "dark matter". Dark matter makes up most of our Universe, but is impossible to detect with a telescope.

Solar System forms 9 billion years after Big Bang

Star bright
The radiant energy pumped out from our Sun keeps life going on Earth. Might other stars also support life? The answer is uncertain. What is certain is that every star has a limited amount of fuel, and when it runs out, the star dies. If the star is heavy enough, it goes out with an explosion called a supernova (left) that burns with the energy of 100 billion Suns.

Global energy

The world is a power-hungry place, but fossil fuels are not replaceable, and burning them releases damaging carbon dioxide (CO_2). The challenge is to find a way of generating energy that is sustainable for the future, without causing harm to our planet.

Energy consumption

World energy consumption is a measure of the total amount of energy consumed by everyone on the planet. However, not all parts of the world generate and consume the same amount of energy. For instance, Chad (in Africa) has the lowest CO_2 emissions per person, since its citizens have limited access to energy. In 2012, the world burned through a total of 560 billion billion joules (J) of energy. The following boxes give a brief account of the energy generation in a few countries in 2012, as well as some of the most important power plants in the world.

Key

 Electricity from fossil fuels

 Electricity from renewable sources

 Electricity from nuclear power

Consumption per 1 million people

CO_2 emissions

1 Mtoe (million tonnes of oil equivalent) = 44,760,000,000 J

MtCO2: Metric tonnes of carbon dioxide released

1 MW = 1,000,000 W
1 W = 1 J per second

Japan

Japan relied mainly on nuclear power for generating electricity until 2011. But that year's earthquake reduced nuclear electricity generation.

 86%   12%   2%

 3 Mtoe   1,238 MtCO2

Kashiwazaki nuclear power plant, Honshu (Peak power: 7,965 MW)
With seven reactors, this is the world's largest nuclear power station, though it was shut down after the 2011 earthquake. To reopen the plant, additional safety measures were planned.

China

China leads the world both in producing electricity from fossil fuels and renewable energy. It is the world's largest emitter of CO_2, but its emissions per person are average.

 77%   21%   2%

 1.8 Mtoe   9,153 MtCO2

Taichung power plant, Taiwan (Peak power: 5,824 MW)
The world's largest coal-fired power station, and also the world's single largest emitter of CO_2.

USA

The USA is the world's second-largest consumer of energy, after China. Thanks to its large oil reserves, the USA is capable of producing nearly 90 per cent of its own energy.

 68%   13%   19%

 7.3 Mtoe   5,287 MtCO2

The Geysers, California (Peak power: 1,517 MW)
More than 350 wells supply hot steam to 22 separate power plants.

Ivanpah Solar Power Facility, California (Peak power: 392 MW)
The world's biggest solar thermal power station, consisting of three power towers and 173,500 mirrors.

Three Gorges Dam, Yichang (Peak power: 22,500 MW)

The world's largest hydroelectric power plant. It has enough steel in it to make 63 Eiffel towers. It generates energy equivalent to that produced by burning 31 million tonnes of coal.

UK

The UK produces more electricity from wind power than from hydropower. With diminishing gas supplies, the country imports more energy than it produces.

68%   13%   19%

3.2 Mtoe   505 MtCO2

London Array (peak power: 630 MW)

The largest offshore wind-power plant in the world. It sits in the sea off the east coast of Britain. Its 175 turbines and two substations are connected together with 210 km (130.5 miles) of high-voltage copper cable.

Drax Power Station (Peak power: 3,960 MW)

Drax Power Station, the largest coal-fired power station in the UK, has been converted so that it can also burn biomass.

How much carbon is emitted

Every manufactured product requires energy to make, and when this energy comes from burning fossil fuels, CO2 and other greenhouse gases are released into the atmosphere. The total amount of greenhouse gases released is called a carbon footprint. This gives us a way to compare the contribution that various activities and products make to global warming.

56 g (1.9 oz) of raw steak (1,000 g)

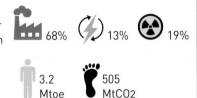

1 kg (2.2 lb) of apples (600 g)

2 kg (4.4 lb) of garden waste (400 g)

1-minute-long mobile phone call (57 g)

3-km (1.8-mile) car journey (739 g)

Watching 28-in TV for 5 hours (380 g)

Heating a saucepan of water (115 g)

Cup of tea (71 g)

CO2 emission in grams

Global electricity generation by energy source

40%
5%
4%
11%
17%
23%

Key
- Coal
- Gas
- Hydroelectric
- Nuclear
- Oil
- Others

Electricity generation accounts for a great deal of energy use. The above chart breaks down the percentage of different energy sources consumed globally in 2012 to generate electricity. An enormous two-thirds of all electricity around the world still comes from fossil fuels. Hydropower is the next largest sector, but accounts for barely one-fifth of all energy production.

Web of energy

Everywhere you look, energy is being transferred from one form into another – from cars and motorbikes whizzing past, burning fuel to release chemical energy, to plants silently converting the Sun's energy to grow. Without energy nothing can happen.

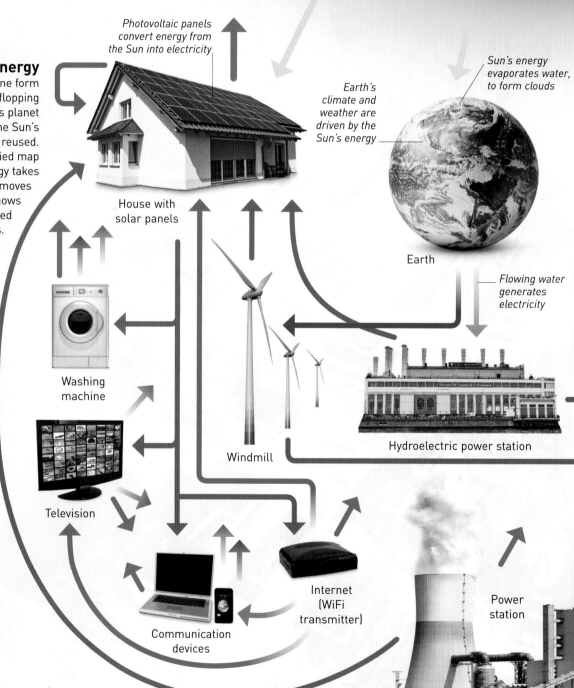

Transmitting energy

Energy transfers from one form to another. These flip-flopping conversions keep this planet bursting with life, as the Sun's energy is used and reused. An energy web is a simplified map that traces the routes energy takes as it changes hands and moves around our world. It shows how things are connected by energy transfers.

Photovoltaic panels convert energy from the Sun into electricity

House with solar panels

Sun's energy evaporates water, to form clouds

Earth's climate and weather are driven by the Sun's energy

Earth

Flowing water generates electricity

Washing machine

Television

Windmill

Hydroelectric power station

Internet (WiFi transmitter)

Communication devices

Power station

Energy key:

Electromagnetic radiation

Heat

Chemical

Electrical

Kinetic

Sound

Gravitational

Radiant

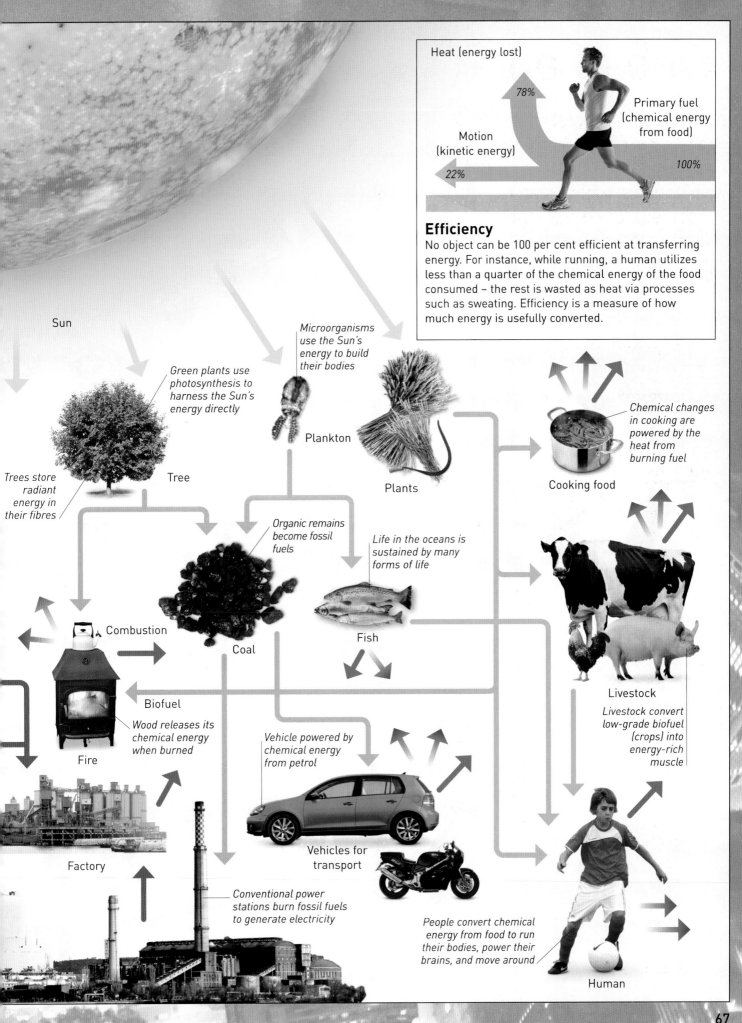

Heat (energy lost)

78%

Motion (kinetic energy)

22%

Primary fuel (chemical energy from food)

100%

Efficiency

No object can be 100 per cent efficient at transferring energy. For instance, while running, a human utilizes less than a quarter of the chemical energy of the food consumed – the rest is wasted as heat via processes such as sweating. Efficiency is a measure of how much energy is usefully converted.

Sun

Microorganisms use the Sun's energy to build their bodies

Green plants use photosynthesis to harness the Sun's energy directly

Plankton

Chemical changes in cooking are powered by the heat from burning fuel

Plants

Cooking food

Trees store radiant energy in their fibres

Tree

Organic remains become fossil fuels

Life in the oceans is sustained by many forms of life

Combustion

Coal

Fish

Livestock

Livestock convert low-grade biofuel (crops) into energy-rich muscle

Biofuel

Wood releases its chemical energy when burned

Fire

Vehicle powered by chemical energy from petrol

Factory

Vehicles for transport

Conventional power stations burn fossil fuels to generate electricity

People convert chemical energy from food to run their bodies, power their brains, and move around

Human

Timeline

Learning how to harness energy has been one of the defining stories in the rise of humans. Each discovery has given us access to more power, and has enabled new technological advances. Understanding energy took longer, but it has been key to our understanding of the Universe.

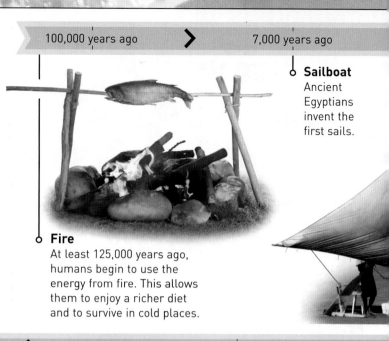

100,000 years ago > **7,000 years ago**

Sailboat
Ancient Egyptians invent the first sails.

Fire
At least 125,000 years ago, humans begin to use the energy from fire. This allows them to enjoy a richer diet and to survive in cold places.

1850 < **1800**

Oil boom (1848)
The world's first oil well is drilled at Baku, Azerbaijan. Oil has much more chemical energy per kilogram than other fuels.

James Joule's calorimeter

Conservation of energy (1837)
Karl Friedrich Mohr declares that energy can be changed from one form to another.

Electromagnetism (1820)
Hans Christian Ørsted discovers link between electricity and magnetism.

Electric motor (1821)
Michael Faraday invents a device to convert electrical energy into kinetic energy.

Energy conversion (1845)
James Joule proves "work" from a falling weight increases the internal energy of water.

Engine efficiency (1824)
Sadi Carnot explains that engines cannot convert 100 per cent of the energy in their fuel.

Faraday disc

Electric battery (1800)
Alessandro Volta invents the first battery.

1900 > **1950**

Replica of Edison's lamp

Wind turbine (1887)
Electricity is first generated from power of wind.

Petrol engine (1886)
Karl Benz makes the first production motorcar.

Light bulb (1879)
Thomas Edison sells first practical electric light bulb to the public.

Greenhouse effect (1908)
Svante Arrhenius explains how Sun's energy keeps planet warm.

Mass-energy (1905)
Albert Einstein shows that mass and energy are really the same thing.

Electron (1897)
J J Thomson discovers the electron.

Electron

Radioactivity (1896)
Henri Becquerel discovers the energy emitted by unstable nuclei.

Turbojet engine (1930)
Frank Whittle invents the turbojet.

Nuclear reactor (1942)
The first functioning nuclear-fission reactor is built in Chicago, USA.

Frank Whittle (right) with a turbojet engine

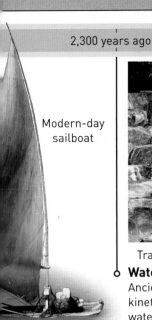

Modern-day sailboat

Traditional water mill

Water wheel
Ancient Greeks use the kinetic energy of running water to turn a wheel.

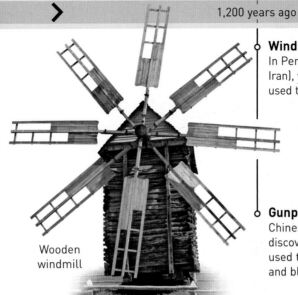

Wooden windmill

Windmill
In Persia (modern-day Iran), windmills are used to grind grain.

Gunpowder
Chinese alchemists discover gunpowder. It is used to fire projectiles and blast rocks.

1750   1700 CE

Heat (1797)
Benjamin Thompson shows there is no limit to the amount of heat that endless mechanical action can generate.

Photosynthesis takes place in green leaves

Photosynthesis (1779)
Dutch scientist Jan Ingenhousz discovers that plants give out oxygen only in sunlight.

Steam engine (1710)
British inventor Thomas Newcomen invents the first steam-powered engine, which is used to pump water out of mines. However, it is inefficient and prone to blowing up. In time, engines revolutionize power delivery.

Replica of Newcomen steam engine

2000   2020

Nuclear bomb (1961)
Tsar Bomba – the largest nuclear bomb ever tested – is detonated.

Cosmic background radiation (1964)
Astronomers Arno Penzias and Robert Wilson accidentally discover evidence for the Big Bang.

Gamma ray bursts (1973)
The most energetic explosions in the Universe are observed for the first time.

Solar panel made of individual PV cells

Solar cell (1963)
Bell Labs in USA builds the first practical solar panels to harvest solar energy.

Laser light show
Laser (1960)
Theodore H Maiman builds the first laser.

Hydroelectric power (2012)
Three Gorges Dam, China – the world's largest hydroelectric power station – opens.

Modern cottage with solar panels on roof

Glossary

ATOM
A tiny particle of matter, made up of protons, neutrons, and electrons.

BIG BANG
The theory that the Universe began in a giant explosion 13.7 billion years ago.

BIOFUEL
A fuel made from, or by, a living thing.

BIOMASS
Material from living things whose stored chemical potential energy can be used to produce biofuel.

CALORIMETER
Scientific equipment used for measuring the calorific value of fuels or food.

Boyle's gas calorimeter

CARBON FOOTPRINT
The total amount of carbon dioxide and other greenhouse gas emissions released by a person, a product, or a nation.

CATALYST
A substance that helps a chemical reaction.

CHARGE
The attribute of matter by which it is affected by electromagnetic forces.

CHEMICAL REACTION
A process in which substances change to become different substances.

COMBUSTION
A rapid chemical reaction in which oxygen combines with a substance, releasing energy. Burning is oxidation accompanied by flames.

CONDUCTOR
A material through which heat or electricity can flow easily.

CONSERVATION OF ENERGY
The theory that states that energy cannot be created or destroyed, but merely changes its form.

ECOSYSTEM
A community of interconnected living things and the environment they live in.

EFFICIENCY
A measure of the amount of useful energy converted by a machine or device.

ELECTRIC CURRENT
The movement of electric charges.

ELECTROMAGNETIC RADIATION
Also known as radiant energy, a form of energy that travels through space at the speed of light.

ELECTROMAGNETIC (EM) SPECTRUM
The range of frequencies of light (EM radiation) from radio waves to gamma rays.

ELECTROMAGNETISM
The forces of attraction and repulsion that occur between electric charges and magnetic fields.

ELECTRON
A tiny, negatively charged particle that orbits the positively charged atomic nucleus.

ENZYME
A natural protein catalyst that makes chemical reactions in the human body happen much more quickly.

FOSSIL FUEL
A fuel made by geological processes from the remains of living things.

FREQUENCY
A measure of the number of times a wave oscillates per second, measured in hertz.

FRICTION
A resistive force produced between two objects when they rub together. It always produces heat and slows down an object down.

GENERATOR
A device for producing electrical energy from kinetic energy.

Nodding donkeys extract crude oil (a type of fossil fuel)

GLOBAL WARMING
The slow but steady rise in average temperatures on our planet.

Faraday disc, the first electric generator, invented in 1821

GRAVITATIONAL POTENTIAL ENERGY
An object's energy in a gravitational field.

GRAVITY
The attractive force felt between objects with mass.

HEAT
The transfer of energy from a hotter to a cooler object, by conduction, convection, or radiation.

HYDROELECTRICITY
The conversion of the gravitational potential energy of water to electrical energy.

INSULATOR
A material through which heat or electricity does not flow easily. Plastics, glass, and the air are examples of good insulators.

Ice lollies melting due to heat in the surrounding air

KINETIC ENERGY
The energy of moving things.

MAGNETIC FIELD
The force field surrounding a magnet or a moving electric charge that causes a magnetic force to act on magnetic materials or moving charges.

MASS-ENERGY
The energy due to an object's mass, as outlined in Einstein's Theory of Relativity.

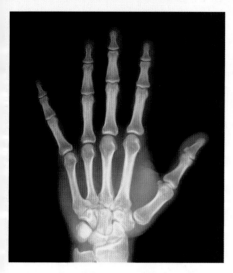

X-ray of a hand

MICROGENERATION
The generation of relatively small amounts of renewable power for individual homes.

NEUTRON
A neutrally charged subatomic particle that helps bind together the nucleus of an atom.

NUCLEAR FISSION
A reaction in which a large atomic nucleus splits into two, releasing energy.

NUCLEAR FUSION
A reaction in which small atomic nuclei join together, or fuse, to form a larger one.

NUCLEUS
The central part of an atom. The nucleus contains protons and, usually, neutrons.

PHOTOSYNTHESIS
The process by which plants convert the Sun's energy into chemical energy, stored in a sugary substance called glucose.

POLLUTION
Introduction of harmful or poisonous substances into an environment.

POTENTIAL ENERGY
Stored energy that – often due to position – is available to use in some form at a later time.

POWER
The rate at which energy is transformed from one form to another.

PROTON
A positively charged subatomic particle found in the nucleus of an atom.

RADIATION
Energy transferred, or radiated, in all directions.

RED GIANT STAR
A star, with a mass similar to our Sun, in the last stages of its life.

RENEWABLE ENERGY
An energy source whose supply is unlimited.

SONAR
A system of detecting objects using sound energy. It is used by ships and submarines.

SMELTING
A process that separates metal from its ore.

SUPERCONDUCTOR
A special material that has zero resistance to an electric current.

SUSTAINABLE
An activity that can be continued without end.

TECTONIC PLATE
Thick, solid slabs of rock that make up the rigid outer layer of Earth.

TEMPERATURE
A measure of internal energy of an object or substance, measured in degrees Celsius, Fahrenheit, or Kelvin.

TURBINE
A machine that converts the linear kinetic energy of water, steam, or gas into rotary kinetic energy.

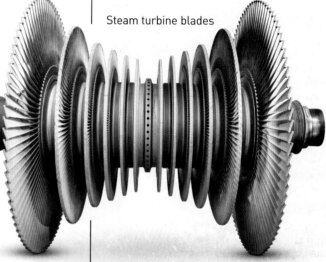

Steam turbine blades

VOLTAGE
A measure of the electric potential difference between two points in a circuit.

WAVE
A way in which energy travels.

WAVELENGTH
The distance between matching parts of a travelling wave.

WORK
Any process that requires energy.

Windmills generate renewable energy

Index

Acknowledgements

Dorling Kindersley would like to thank: Virien Chopra, Rupa Rao, and Deeksha Saikia for editorial assistance; Nidhi Rastogi for design help; Helen Peters for indexing; and Deepak Negi for picture research.

The publisher would like to thank the following for their kind permission to reproduce their photographs: (Key: a-above; b-below/bottom; c-centre; f-far; l-left; r-right; t-top)

2 Alamy Images: Accent Alaska.com (t). Dorling Kindersley: The Science Museum, London (bl). Dreamstime.com: Rumos (tl). Getty Images: subtik / E+ (cl). iStockphoto.com: 4FR (cl). 2–3 NASA: (c). 4 Corbis: 13 / Atomic Imagery / Ocean (cr); RG Images / Stock4B (cl). Getty Images: Antonio Iacobelli / Moment (tl); Miguel Navarro / Stone (bl); Mike Powell / Allsport Concepts (c). SuperStock: Photri Images (br). 5 Getty Images: Dimitri Vervitsiotis / Photographer's Choice RF (cr). 6–7 Dreamstime.com: Farek (c). 6 NASA: (clb). 7 Dreamstime.com: Ashwin Kharidehal Abhiram (crb); Andrey Kotko (tr). Getty Images: subtik / E+ (cl). Volvo Car Group: (bc). 8 Dreamstime.com: Paul Prescott (tl). 9 Getty Images: Anna Gorin / Moment (cr). Science Photo Library: Peter Matulavich (br). 10 Corbis: Lloyd Cluff (tl); Martin Harvey (bl). Getty Images: JTB Photo (cl). 10–11 Dreamstime.com: Adambowers. 11 Dreamstime.com: Zacarias Pereira Da Mata (crb). Science Photo Library: Mike Agliolo (tr). 12 Corbis: Martyn Goddard (tr); Neumann & Rodtmann (cl). Getty Images: Harry Engels (bl). 12–13 Science Photo Library: Ted Kinsman (c). 13 Corbis: 13 / Atomic Imagery / Ocean (br); PCN (cr). 14 Corbis: Hulton-Deutsch / Hulton-Deutsch Collection (crb). 15 Berkeley Lab: Roy Kaltschmidt (cr). naturepl.com: Doug Perrine (cr). Science Photo Library: Alex Hyde (t). 16 Dreamstime.com: Akulamatiau (bl); Emanoo (tl). Getty Images: Mike Powell / Allsport Concepts (r). 17 Alamy Images: Kenneth R. Whitten / Design Pics Inc (r). Corbis: Sebastian Kaulitzki (bc). 18 Corbis: (tc). 18–19 Alamy Images: laurentiu iordache (c). 19 Alamy Images: Accent Alaska.com (t). Dreamstime.

com: Rumos (tll); Sippakorn Yamkasikorn (cl). Getty Images: Rob Melnychuk / Digital Vision (cla); Flip Nicklin / Minden Pictures (tr). 20–21 Getty Images: Dimitri Vervitsiotis / Photographer's Choice RF (r). 20 Corbis: Stefano Bianchetti (bc). Dreamstime.com: Gadagj (bl). 21 Alamy Images: Jerry Choi (tc). Getty Images: Kristina Greke / E+ (bl). NASA. 22 Getty Images: Bloomberg (c); Pete Starman / Photographer's Choice RF (br). iStockphoto.com: 4FR (clb). 22–23 Alamy Images: Don Farrall / Stone (c). 23 OceanwideImages.com: (c). Science Photo Library: R. Maisonneuve, Publiphoto Diffusion (br). 24 Photoshot: ISRO / Xinhua (l). 25 Corbis: Bo Bridges (c); Michel Gunther / Photononstop (tr). Ed Hengeveld: (l). NASA: (bc). 26 Dreamstime.com: Tetiana Novikova (c). Getty Images: Duane Walker / Moment (l). 27 Corbis: RG Images / Stock4B (tr). East West Dive and Salvage: (br). Science Photo Library: Charles D. Winters (cl). 28 Dorling Kindersley: The Science Museum, London (cb). The Library of Congress, Washington DC: Duplessis / Joseph-Siffrède (cla). 28–29 Dreamstime.com: Richard Valdez (c). 29 Dreamstime.com: Phadventure (cr); Yourthstock (tr). Getty Images: Bloomberg (br). 30 Corbis: Yi Lu / Viewstock (br); Paul Rapson (bl). 31 Alamy Images: David R. Frazier Photolibrary, Inc. (br); Paul Rapson (bl). Corbis: (tl). 32–33 SuperStock: Photri Images (c). 32 The Library of Congress, Washington DC: Turner / Orren Jack (bl). 33 Getty Images: Shone (c). NASA: (tr). 34–35 Dreamstime.com: AntonBalazh (bc). 34 Corbis: De Agostini Picture Library (cl). Science Photo Library: CERN (tl). 35 Alamy Images: Oliver Furrer (crb). Science Photo Library: Antonio Iacobelli / Moment (tl). Robert Harding Picture Library: Jen Judge (tl). 36–37 Corbis: Nigel Pavitt / JAI (c). 36 Alamy Images: Joseph Sohm / Visions of America, LLC (cra). Dorling Kindersley: The National Railway Museum, York / Science Museum Group (bl). Dreamstime.com: Digitalpress (tl). Science Photo Library: Sheila Terry (t). 38 Getty Images: Danita Delimont / Gallo Images (tll); Handout (c). 38–39 Dreamstime.com: Khunaspix (b). 39 Getty Images: Images Etc Ltd (tl); Mark Thiessen / National Geographic Magazines (cl). 40 Alamy Images: Werner Forman Archive /

Heritage Image Partnership Ltd (tll). Getty Images: Science & Society Picture Library (bl). 40–41 SEIJI CHIBA: (bl). 41 Toyota Motor Europe: (tr). 42 Dorling Kindersley: HPS Museum of Leeds University (cra). Getty Images: Blackstation / Moment Open (bl); Library of Congress / digital version by Science Faction (cb). 42–43 Masterfile: (b). 44 123RF.com: Piotr Pawinski (bl). Alamy Images: Phanie / RGB Ventures / SuperStock (bc). 44–45 Dreamstime.com: Alexsalcedo (tc). 45 Alamy Images: Jeff Morgan 15 (b). Dreamstime.com: Suljo (tl). Getty Images: Stephen Barnes / Technology (tc). Dorling Kindersley: The Science Museum, London (cl). Getty Images: Caiaimage / Chris Ryan (tr). 46–47 Getty Images: AFP (cr). 47 Corbis: BURGER / phanie / Phanie Sarl (bl). Getty Images: TiaClara (cl). Science Photo Library: Robert Brook (t). 48 Corbis: Hulton-Deutsch Collection (bl); Ken Seet (tl). Getty Images: Paul J. Richards / AFP (cr). Science & Society Picture Library (cr). 49 Dreamstime.com: Photozirka (tl). Getty Images: Ariel Skelley / Blend Images (cra). Science Photo Library: Dr Seth Shostak (b). 50 Getty Images: Peter Macdiarmid (tl). 50–51 Getty Images: Miguel Navarro / Stone (c). 51 Altaeros Energies: (tr). Artwork by Luke Jerram.: (br). Getty Images: The Image Bank (tc). 52–53 Getty Images: Laurent Gillieron / AFP (tl). 52 Corbis: Stringer Shanghai / Reuters (bl). Dreamstime.com: Radha Karuppannan (cra). 53 Bridgeman Images: Galleria degli Uffizi, Florence, Italy (cl). Photoshot: Construction Photography (b). 54 Alamy Images: Egmont Strigl / age fotostock (tl). Getty Images: UniversalImagesGroup (tl). 54–55 Corbis: Cheng Min / Xinhua Press (b). 55 Alamy Images: Simon Waldman (tr). Corbis: Kim Hart / Robert Harding World Imagery (tl). 56 Alamy Images: Peter Bennett / Danita Delimont (bl). Rex Features: Global Warming Images (tl); Sipa Press (b). 57 Alamy Images: Brian J. Skerry / National Geographic Image Collection (t). Emmanuel LATTES (bl). Rex Features: Geoffrey Swaine (r). 58 Alamy Images: Marius Graf (tr). NASA: (bl). 58–59 Tom Chance: (b). 59 Corbis: 68 / Don Mason / Ocean (cr). Dreamstime.com: Peter Wollinga (b). 60 Alamy Images: Ashley Cooper pics (tl). Getty Images: Yoshikazu Tsuno / AFP (cl, bl). 60–61 Getty Images: Maximilian Stock Ltd (c). 61 BMW Group: (tl). VINCENT CALLEBAUT ARCHITECTURES: (cr). 62–63 NASA: ESA / Herschel / PACS / MESS Key Programme Supernova Remnant Team / Allison Loll / Jeff Hester (Arizona State

University) (bc). 62 Corbis: Roger Ressmeyer (bl). NASA: ESA; G. Illingworth, D. Magee, and P. Oesch, University of California, Santa Cruz; R. Bouwens, Leiden University; and the HUDF09 Team (c). Science Photo Library: Take 27 Ltd. (c). 63 NASA: (tl). Science Photo Library: Harald Ritsch (c). 64 Alamy Images: Aerial Archives (c). Corbis: Steve Proehl / Proehl Studios (bl). Getty Images: The Asahi Shimbun (bl); michaeliao27 / Moment Open (bc). 64–65 Corbis: Cameron Davidson (Background). 65 Alamy Images: adrian arbib (bc). Getty Images: ChinaFotoPress (tl); Chris Ratcliffe / Bloomberg (cb). 66–67 Corbis: Cameron Davidson (Background). Dreamstime.com: Dimitar Marinov (tl). 66 Dreamstime.com: Dvad (cr); Grzym (cb). 67 123RF.com: Gui Yongnian (clb). Dorling Kindersley: Wayne Tolson (bc). Dreamstime.com: 3quarks (ca); Maksim Toome / Mtoome (cb). 68–69 Corbis: Cameron Davidson (Background). Dreamstime.com: Coplandj (tl). 68 Corbis: Hulton-Deutsch Collection (br). Dorling Kindersley: The Science Museum, London (c, bl). Fotolia: valdis torms (bc). Getty Images: altrendo images / Stockbyte (tl); Science & Society Picture Library (cr). 69 Corbis: Du Huaju / Xinhua Press (br). Dreamstime.com: Byelikova (tr); Carroteater (tc); Suttipong Sutiratanachai (bl); Dmitry Kalinovsky (bc). Getty Images: De Agostini Picture Library (cra). 70 Dorling Kindersley: The Science Museum, London (ca, tr). Dreamstime.com: Wenbin Yu (bl). 70–71 Corbis: Cameron Davidson (Background). 71 Alamy Images: Monty Rakusen / Cultura Creative (RF) (cr). Dreamstime.com: Itsmejust (tl). Getty Images: Sergio Pitamitz / National Geographic Magazines (b)

Wallchart: Alamy Images: Stephen Barnes / Technology cr; laurentiu iordache cla/ (flame), Brian J. Skerry / National Geographic Image Collection bc, Kenneth R. Whitten / Design Pics Inc tr; Tom Chance: cra/ (House); Dorling Kindersley: The Science Museum, London cb/ (glass globe); Dreamstime.com: Farek tl, Khunaspix crb; Getty Images: Don Farrall / Stone cb, Miguel Navarro / Stone bl, Paul J. Richards fcrb, Universal Images Group cl, cb; Science Photo Library: Alex Hyde cra, Ted Kinsman cla, Maximilian Stock Ltd br; Toyota Motor Europe: cr/ (Yaris)

All other images © Dorling Kindersley

For further information see:
www.dkimages.com